# istry and
# gy of Coal

# The Chemistry and Technology of Coal

*Editor*

**Neeraj Singh**

scitus academics

**The Chemistry and Technology of Coal**
Edited by **Neeraj Singh**

ISBN: 978-1-68117-681-9
Library of Congress Control Number: 2016958838


616, Corporate Way, Suite 2, 4766,
Valley Cottage, NY 10989

www.scitusacademics.com



**Notice**



Printed in United States of America on Acid Free Paper

# Preface

Coal is a natural solid fuel, which is among the most plentiful and significant sources of energy in the world. From its introduction four thousand years ago as a fuel for cooking and heating to its use in the production of electricity in the 19th and 20th centuries and as a chemical feedstock, coal and natural gas and oil have remained an important supply of electricity. Chemists had little knowledge of the molecular structure and composition of the various types of coal until the last 100 years, and until the 1920s they still believed that coal consisted of carbon mixed with hydrogen containing impurities. Their two methods of analyzing or possibly separating coal into its components, destructive distillation (heating from contact with solvent extraction and air), showed only that coal contained substantial carbon and smaller proportions of the hydrogen, nitrogen, oxygen and sulfur elements. Ash is constituted by inorganic compounds such as aluminum and silicone oxides.

This state of the art compilation provides a comprehensive and comprehensive overview of the chemistry and technology of coal in the 21st century and also extends the relationship between processes in the coal industry and environmental regulations and the effects of combustion products on the environment. The book will provide valuable information resources for scientists and engineers working in the coal and catalyst manufacturing industry to help and update clean coal technology and expert engineering and chemistry researchers and practitioners.

Coal is most likely the dirtiest of fossil fuels. When burnt, it produces emissions that contribute to climate change, causes acid rain and pollutes water. But coal is not a soothing remnant of the Industrial Revolution-it generates the majority of the world's electricity and will probably continue to do so if it is a rich and cheap resource. Clean coal technology aims at reducing harsh environmental effects by using a number of technologies to clean coal and reduce emissions.

# Table of Contents

# Chapter 1

# Evaluation of Pneumatic Inclined Deck Separator for High-Ash Indian Coals

**Nikhil Gupta***

*Postdoctoral Research Associate, Department of Mining and Minerals Engineering, Virginia Tech, Blacksburg, VA 24061, USA*

## ABSTRACT

Application of pneumatic separators in coal beneficiation is increasing rapidly over the last decade primarily due to their low capital and operating costs, and waste handling problems associated with traditional wet processing methods. Large amount of shale/rock that is extracted in coal production can be removed prior to transportation at the mine face by using this methodology. Due to the limited washing facilities in India, most of the thermal power plants burn raw coal from run-of-mine (ROM) to generate electricity. This practice causes poor utilization efficiency, high operating and maintenance costs, and high emission rates for the power plants. One potential method that can be utilized is the air-fluidized inclined vibrating deck technology. The technology was demonstrated on a pilot-scale at different coal washeries in India at a feed rate of 5-ton per hour. The pilot-scale evaluation showed that 20 %–25 % high-ash incombustible material can be eliminated from ROM feed with only minor losses in energy content (<10 %) from respective ROM coal. Furthermore, a feasibility analysis showed significant economic gains in terms of transportation cost, improving power-plant efficiency, and reducing emissions rates by using the technology.

**Keywords:** Coal deshaling Pneumatic separationInclined deck Vibrating table Fluidization

# 1 INTRODUCTION

In 2014, the United Nations' Intergovernmental Panel on Climate Change (IPCC) issued the Fifth Assessment Reports (AR5). The report from the Working Group I (The Physical Sciences Base) stated that the warming trend which began since the mid-twentieth century is real and it is "*very likely*" due to the increase in the atmospheric concentrations of the anthropogenic greenhouse gases (GHGs) such as carbon dioxide ($CO_2$), methane ($CH_4$), nitrous oxide ($N_2O$) and others. Of these, $CO_2$ accounts for 65 % of the greenhouse gases generated largely from burning fossil fuels and other industrial applications. The atmospheric $CO_2$ concentration of 390 ppm in 2014 far exceeded the range (180–300 ppm) observed during the last 650,000 years, and the use of fossil fuels was its major source. The report clearly suggests that there is a dire need to reduce the GHG emissions to the level that can be absorbed by the Earth's system (IPCC 2013).

India is the third largest GHGs emitter in the world, behind China and the United States (World Bank 2015), and its energy supply is largely dependent on coal. As one of the fastest growing economies of the world, its coal consumption is expected to increase rapidly, along with all of the environmental concerns associated with burning coal. An attractive first step toward reducing the $CO_2$ emissions from the coal-fired power plants in India would be to increase the availability of beneficiated coals by using appropriate separation technologies. It has been shown that use of beneficiated coals can increase thermal efficiencies by 2 %–3 % on existing PC boilers, possibly as much as 4 %–5 %, and can thereby reduce $CO_2$ emissions up to 15 % (Zamuda and Sharpe 2007). If the low-ash coals are used as feedstocks for advanced clean coal technologies, such as Integrated Gasification Combined Cycle (IGCC) or supercritical Pulverized Coal Technology (PCT), the thermal efficiencies can become as high as 45 %, which will further reduce the GHG emissions. The use of beneficiated coals will also have many other positive benefits. For example, beneficiated coal can increase in the plant utilization factor (PUF) by 13 %, minimize fouling and slagging problems, reduce fugitive dusts, and decrease in the amounts of unburned carbon in fly ash (Couch 2002). More importantly, shipping beneficiated coals will increase railway capacities on an equivalent heat basis, which will be of paramount importance in India.

In the fiscal year 2013–2014, India consumed 739.7 million metric tons (MMT) of coal, out of which 71 % was used for thermal power plants (MOP 2015). Of this, domestic supply could only fulfill 76 % of the country's demand. Unfortunately, the coal industry in India has been slower than in other countries in embracing coal beneficiation as part of their normal operating practice. At present, domestic coal accounted only less than 20 % was beneficiated, out of which 97 % operations involve water based separation methods (MOC 2015). The reasons for minimal beneficiation includes the difficulty in cleaning Indian coals, the high capital and operating costs associated with the conventional wet-beneficiation processes, and the lack of capital. Ash-forming minerals are finely dispersed in the Indian coals throughout the carbonaceous matrix, making them difficult to be removed using conventional physical separation methods. Further, water used in traditional water-based beneficiation plants is a scarce resource, especially in the coal mining areas. As per an estimate, the water table in India is shrinking rapidly (Tiwari et al. 2009), which makes alternative methods such as dry coal cleaning extremely beneficial.

Preliminary studies conducted for dry coal beneficiation suggest that these methods can be an effective techno-economical tool for the Indian coal industry (Honaker 2007). Significant amount of rock are extracted in order to recover the coal; reportedly resulting in 60 %–70 % of the raw material as waste. The haulage, processing, and combustion of rock represent a significant loss of limited available resources and create negative environmental impact (Biswal et al. 2002). Removing unwanted rock from run-of-mine (ROM) coal is referred to as *deshaling*, which normally involves separation of high-density rock from raw coal in a density-based process. In contrast to traditional coal preparation plants, the separation relative density (*RD*) for deshaling is higher with a typical target of *RD* = 2.0 or greater (Honaker et al. 2008).

Several dry coal separation technologies, such as, All-Air® Jig (allmineral GmbH), Air Dense Medium Fluidizied Bed (ADFMB), Tirbo-electrostatic Separators, Rare Earth Magnetic Separators (REMS), and X-ray Sorting technologies, were proven effective in terms of separation to achieve low-ash products for coarse to mid-size fractions (Dwari and Rao 2008; Zhao et al. 2014). In previous years, All-Air® Jig (ISR 2005; Gouri et al. 2007) and ADFMB (Sahu et al. 2005) have been tested for coarse Indian coal feedstock, and the data showed moderate recoveries to achieve low-ash product. The reason for lower recoveries could be due the poor formation of Indian coals, where particles are not liberated well at the coarse size fraction. However, these technologies can be applicable for the purpose of deshaling to remove high ash rock from ROM coals as pre-beneficiation method. Using dry coal separators for pre-beneficiation can provide significant economic savings. In general, the most cost-effective approach is to install the dry deshaling unit as near as possible to the mine extraction face to reduce transportation and maintenance costs (Honaker et al. 2006).

One such emerging technology that has been proven very effective in major coal producing countries is the air-fluidized inclined vibrating deck technology.

The method utilizes integrated effect of vibration force (maceration) and continuous air flow for segregation of particles according to their densities (refer Fig. 1). The present communication highlights pilot-scale evaluation of the vibrating air-deck technology tested at the different coal beneficiation plants in India for the specific purpose of deshaling. In addition, a feasibility study was conducted using a cost-benefit model described by Gupta et al. (2014) to demonstrate economic gains by using this technology.

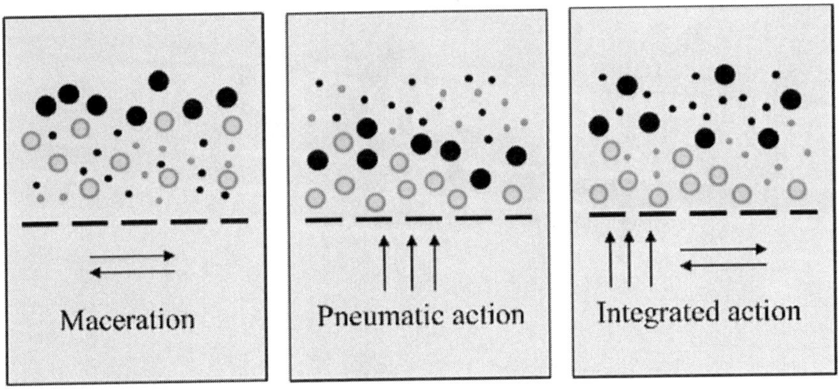


**Figure 1.** Schematic shows operating principle of inclined deck separator

# 2 TESTING PROCEDURE

## 2.1 INCLINED DECK SEPARATOR TECHNOLOGY

The dry coal vibrating air-table deshaling technique utilizes density separation in autogenous medium using a shaking table concentrator. The air table generates a helical motion with air stratification (as illustrated in Fig. 2a) and generates multiple products of clean coal, middlings, and reject (Dwari and Rao 2007). The separating compartment comprises a table with a perforated deck; vibrating motors, dust collecting air-chamber, and table support/adjusting pulleys. The main centrifugal fan provides continuous flow of air that passes upward through the perforated deck at a sufficient rate to transport and fluidize the particles. Diagonal riffles on the vibrating deck move material towards a back plate. The deck width is reduced from the feed to refuse discharge end. Furthermore, two dust collection systems are installed to clean the recycled air, and to remove the coal dust from air being emitted into the atmosphere.

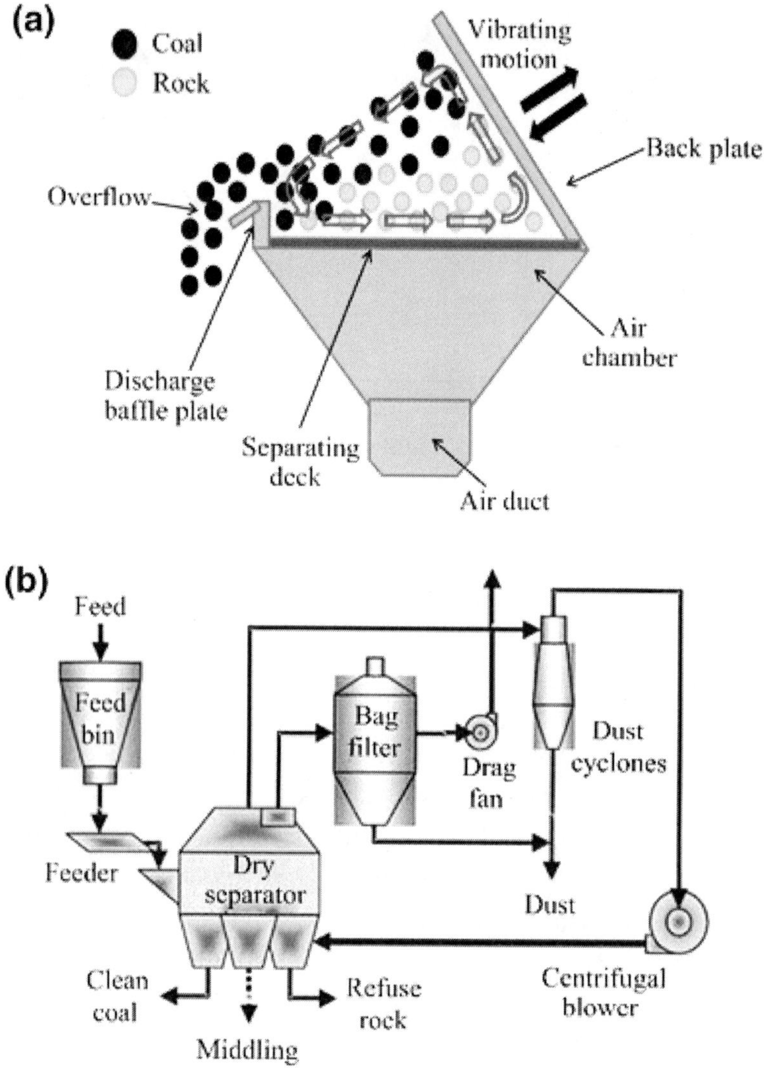


**Figure 2**. Schematics showing **a** working mechanism and **b** process flow diagram of vibrating air-table separators (*after* Gupta 2011)

The bulk of air used for fluidization is recycled back to the table (~90 %) after being cleaned in a dust cyclone, while the rest 10 % is emitted to the atmosphere after passing through a bag-house filter. The bag-house filter creates a negative pressure within the unit to trap all the dust generated by the process. A simplified process flowsheet is shown in Fig. 2b.

The feed bin vibrating feeder controls the feed rate to the separating table. As ROM coal is introduced to the separation zone, a thick bed of particles forms on the vibrating deck. The heavier or high-density particles transport through the bed, comes in contact with the deck, and move toward the back plate under the influence of the inertia force generated due to deck vibration. Particles when strike the back plate, they move inward and upward in the direction of the discharge side of the table. During operation, lighter or low-density particles float on the top of the bed as they are pushed up at the back plate, and flow to the discharge point. As such, the lighter particles create an upper layer that is collected alongside the deck length. Particles of sufficient high-density sinks through the autogenous medium formed due to the bed of particles fluidized due to constant air-flow, and report back to the deck surface. These high-density particles are then forced by both vibration and the continuous influx of new feed material to transport in a helical pattern at the narrow end of the deck where the final reject is collected. Two or three products are collected from the front side of the table. For pilot-scale evaluation, a 5-tph air-table deshaling unit utilizing a 1 $m^2$ air table was tested at three different test sites in India. An immanent advantage of this technology is that it is a dry separator, which delivers a dry product without incurring any dewatering costs.

## 2.2 Test sites description

Pilot-scale testing sessions were conducted at the two private coal groups, Aryan Energy Pvt. Ltd. in Talcher and Bhushan Power & Steel Ltd. in Jharsuguda, and at the Kargali Washery, Coal India Ltd. located in Bokaro, Jharkhand. The coals procured for evaluation at the Aryan Energy site were from two different seams, Balaram and Hingula, of the Mahanadi Coalfields. These coals have a relatively high feed ash and are generally difficult to wash due to high proportions of near-gravity material (NGM). Similarly, the coals received for assessment at the

Bhushan Power & Steel complex were mixed of Mahanadi and IB-Valley coalfields. The facility process coals from multiple sources. At the Kargali Washery, the ROM coal received for evaluation was from three different coal seams of the Karanpura coalfields. These coals are also contains high ash with a high percentage of NGMs. For the coal samples evaluated, average feed ash contents were reported 48.7 %, 54.9 %, and 54.6 % on a dry-basis (db) for the Aryan, Bhushan, and Kargali facility, respectively. Typical surface moisture is ranged as low as 3 %–7 % and as high as 13 %–17 % during rainy season.

## 3 RESULTS AND DISCUSSION

The pilot-scale studies in India were conducted in a statistical designed experimental set-up for the three test sites. The detail description of experimental design has been described elsewhere (Gupta et al. 2016). The major operating parameters were adjusted in each pilot test run. Once the unit achieved steady state, and the samples for clean coal, middlings, and reject were taken in three increments. The operating variables evaluated during the pilot-scale testing in India were the table slope (length-wise angle), the table oscillating frequency, and the main fan motor speed (Honaker and Luttrell 2007; Gupta et al. 2012). Honaker et al. (2008) parametric studies on the pilot-scale unit for the U.S. coals and indicated that the table cross-wise angle had minimal influence on the separation performance, and therefore, can be held constant.

Table 1 exhibits a comprehensive analysis of the products obtained from the pilot tests conducted at the three test sites. This analysis includes product mass (wt%), reject mass (wt%), reject ash content (wt%), the energy loss (kcal/kg %) relative to the feed, organic matter recovery (wt%), and mineral matter rejection (wt%). Mathematically, percent energy loss can be calculated from the three-product formula utilizing the energy contents (kcal/kg) of feed, clean coal, and the reject streams (Wills and Napier-Munn 2006) from the following equation:

**Table 1.** Results from pilot-scale tests for the three coal production facilities

| Location | Test no. | Product mass (Clean Midd.) (wt%) | Reject mass (wt%) | Reject ash (wt%, db) | Energy loss (kcal/kg, %) | Organic matter recovery (wt%) | Mineral matter rejection (wt%) |
|---|---|---|---|---|---|---|---|
| Aryan energy (Avg. feed ash, 48.7 %, db) | 1 | 85.2 | 14.8 | 67.7 | 9.0 | 92.2 | 21.2 |
| | 2 | 82.3 | 17.7 | 60.9 | 13.2 | 87.8 | 22.9 |
| | 3 | 75.6 | 24.4 | 60.4 | 18.9 | 82.3 | 30.2 |
| | 4 | 85.0 | 15.0 | 58.9 | 11.6 | 89.1 | 18.7 |
| | 5 | 79.6 | 20.4 | 64.1 | 14.2 | 87.1 | 26.9 |
| | 6 | 89.5 | 10.5 | 64.6 | 7.3 | 93.4 | 13.7 |
| | 7 | 88.1 | 11.9 | 61.6 | 8.5 | 92.2 | 15.9 |
| | 8 | 93.2 | 6.8 | 62.2 | 4.8 | 95.6 | 9.1 |
| | 9 | 76.1 | 23.9 | 61.5 | 17.7 | 83.6 | 30.7 |
| | 10 | 80.0 | 20.0 | 63.1 | 13.4 | 87.9 | 28.1 |
| | 11 | 65.8 | 34.2 | 63.4 | 24.8 | 77.2 | 43.6 |
| | 12 | 79.5 | 20.5 | 66.3 | 13.3 | 88.2 | 28.2 |
| | 13 | 76.9 | 23.1 | 66.8 | 14.7 | 87.1 | 32.4 |
| | Avg. | 81.3 | 18.7 | 63.2 | 13.4 | 87.8 | 24.8 |
| Bhushan complex (Avg. feed ash, 54.9 %, db) | 1 | 84.1 | 15.9 | 72.6 | 10.0 | 91.6 | 20.4 |
| | 2 | 85.5 | 14.5 | 72.6 | 9.0 | 92.4 | 19.0 |
| | 3 | 86.4 | 13.6 | 70.7 | 8.4 | 92.9 | 18.3 |
| | 4 | 81.7 | 18.3 | 70.2 | 11.8 | 89.8 | 23.8 |
| | 5 | 83.4 | 16.6 | 69.2 | 11.1 | 90.2 | 21.2 |
| | 6 | 84.1 | 15.9 | 73.4 | 9.6 | 92.1 | 20.8 |
| | 7 | 73.5 | 26.5 | 71.4 | 18.7 | 83.5 | 31.9 |
| | 8 | 77.8 | 22.2 | 73.0 | 14.2 | 88.0 | 28.1 |
| | 9 | 80.9 | 19.1 | 70.2 | 11.8 | 89.9 | 25.9 |
| | 10 | 86.6 | 13.4 | 77.4 | 6.3 | 95.3 | 19.9 |
| | 11 | 79.4 | 20.6 | 70.5 | 14.2 | 87.5 | 25.4 |
| | 12 | 73.5 | 26.5 | 66.4 | 11.4 | 82.4 | 32.7 |
| | 13 | 84.0 | 16.0 | 64.0 | 12.2 | 88.7 | 19.5 |
| | Avg. | 81.6 | 18.4 | 70.9 | 11.6 | 90.0 | 23.8 |
| Kargali washery (Avg. feed ash, 54.6 %, db) | 1 | 72.2 | 27.8 | 83.4 | 10.4 | 94.1 | 41.6 |
| | 2 | 83.8 | 16.2 | 79.0 | 6.6 | 95.4 | 26.2 |
| | 3 | 72.8 | 27.2 | 77.9 | 14.1 | 89.4 | 36.9 |
| | 4 | 79.3 | 20.7 | 85.9 | 7.4 | 96.6 | 29.3 |
| | 5 | 83.8 | 16.2 | 85.0 | 5.6 | 97.2 | 24.3 |
| | 6 | 74.9 | 25.1 | 71.1 | 14.3 | 88.0 | 36.2 |
| | 7 | 74.9 | 25.1 | 82.9 | 10.2 | 93.9 | 35.9 |
| | 8 | 83.4 | 16.6 | 81.6 | 6.1 | 96.2 | 27.3 |
| | 9 | 68.5 | 31.5 | 74.5 | 18.7 | 84.9 | 40.6 |
| | 10 | 87.4 | 12.6 | 83.1 | 4.8 | 97.2 | 18.9 |
| | 11 | 79.9 | 20.1 | 78.8 | 9.0 | 93.7 | 30.1 |
| | 12 | 78.8 | 21.2 | 76.6 | 10.5 | 92.0 | 30.6 |
| | 13 | 75.3 | 24.7 | 79.3 | 11.2 | 92.2 | 36.3 |
| | Avg. | 78.1 | 21.9 | 80.0 | 9.5 | 93.5 | 32.1 |

$$\%\left(\frac{\text{kcal}}{\text{kg}}\right)\text{Loss} = \%\text{Reject mass} \times \frac{\left(\frac{\text{kcal}}{\text{kg}}\right)\text{ of reject}}{\left(\frac{\text{kcal}}{\text{kg}}\right)\text{ of feed}}$$

In addition, the analysis includes organic matter recovery of product (clean coal + middling) stream obtained in each test. This indicator represents amount of "organics" recovered in the product relative to the feed, which can be estimated from the mineral matter (inorganic) content of each product. Similarly, mineral matter rejection is represents the amount of inorganics in reject stream relative to feed.

The testing sessions at the Aryan Energy, Bhushan Power & Steel Complex, and Kargali Washery sites successfully demonstrated that the dry coal inclined deck deshaling technology was capable of removing high-ash impurities with modest energy losses from the raw coal presently being processed by these coal groups. From the analysis, on an average 18 %–22 % of the waste (high-ash rock) material can be removed from the ROM coals with minimal loss (9.5–13.4 kcal/kg, %) of energy content from feed to the reject stream. In other words, 1 in every 4–5 rail wagons (full-load) can be eliminated from ROM coal prior shipping with only a minor reduction in heat content by using the low-cost dry deshaling technology.

For Aryan and Bhushan coal samples, the vibrating air-table technology was able to achieve organic matter recovery on an average 87.8 %–90 % of all tests, while the average mineral matter rejections were about 24 % only. In some tests, higher rejections were reported (e.g. test 11 for Aryan) but at the expense of significant loss in heat content of coal in reject. The low mineral matter rejection data for Aryan and Bhushan suggest that the ROM coal carry significant amount of middlings material than the pure high-ash rock material. As such, further liberation of particles in these samples is necessary for better separation of ash from coal.

However, significant improvements were seen for the case of Kargali coal samples using the deshaling technology. In most of the tests, more than 20 % of the ROM material of ash content 80 % was rejected with <10 % loss in heat content of coal. The separation was much better as evident from high organic matter recoveries in most test trials. For the operating conditions in test 1, the organic recovery was recorded 94.1 % with the highest mineral matter rejection of 41.6 %. The results suggest that the Kargali coal sample carry large proportion of pure rock material in ROM coal, and would be a better candidate for deshaling.

Therefore, the pre-combustion removal of these high-ash impurities increased coal heating value and reduced ash disposal and coal transportation requirements without introducing new issues associated with traditional wet coal processing systems (such as, slurry disposal, water pollution, etc.). These analyses suggest that overall transportation costs may be reduced by up to 30 % when compared on an equivalent energy content basis. This value is significant when one considers the fact that many of the railing distances can exceed 1200 km over rail lines in India that are typically overloaded. Recently, as per the Indian Ministry of Environment & Forest, from June 2016 it will be mandatory for producers to supply coal not more than 34 % ash content, if transported to more than 500 km or in environmentally sensitive areas. As such, implementing low-cost vibrating air-table separators for deshaling purpose will be highly beneficial to the ever-growing Indian coal industry.

### 3.1 Economic and efficiency analysis

Using the pilot-scale test results for each site, a feasibility analysis was conducted to demonstrate impact the dry deshaling separation technology on power generation market. For the purpose, cost-benefit model developed by Gupta et al. (2014) was utilized. The model is specifically designed for the Indian coal market from a coal-fired power industry perspective. This model includes variable cost factors, such as, transportation cost (equivalent energy content) as a function of distance, ash and moisture handling cost, and other miscellaneous costs. For analyses, these factors were kept constant, that is, the transportation distance was assumed 200 km, and the ash and moisture handling cost were kept at US $2.5/ton of water and US $ 1/ton of ash, respectively. The target cost of thermal power plant boilers was kept constant to US ¢1.36/kW-h identified by the researchers from extensive studies (Gupta et al. 2014). Finally, the moisture ash-free (maf) energy content of coal was assumed to 7500 kcal/kg based on preliminary analysis obtained for the three coal samples.

The aforementioned model evaluates thermal power plant efficiency with variation in ash content of the material to be burned. For the purpose, the model incorporates performance data obtained by Bhatt (2006), which examined various power plants in the range of 30–500 MW at a variable feed ash material ranging from 6 %–75 % ash, has been utilized. The maximum theoretical efficiency for the thermal power plants is 45 %. The study determined baseline efficiency to be 33.5 % for the Indian power plants (Gupta et al. 2014) using the de-rating adjustment of 74.5 % for 500 MW. In addition, various efficiency corrections factors, such as, boiler, auxiliary, and steam, were defined as a function of ash content in the model using the performance data reported in the literature (Bhatt 2006). Furthermore, the model also accounted efficiency losses due to moisture as 0.15 % per 1 % surface moisture in the power plant. For calculations, the moisture was assumed constant to 8 %.

Figure 3 demonstrate an example of feasibility study under the specified economic parameters for the results obtained for one of the pilot-tests conducted with Kargali coal sample. Rows '*a–c*' indicated direct data obtained from the tests, while rows '*d–e*' shows separation indicators. As described earlier, row '*f*' represents de-rated power plant efficiency, which is further adjusted with correction factors (rows '*f–k*') to calculated net thermal efficiency as shown in row '*j*'. Using this net efficiency and energy content of each stream (row '*c*'), actual energy content to be utilized in power plant was calculated that can be converted to gross worth ($/ton) of material (see row '*o*'). After adjusting with relevant cost factors (rows '*p–s*'), the net worth of each product was calculated (in row '*t*').

| | COST ANALYSIS | | Product | Refuse | Feed |
|---|---|---|---|---|---|
| *a* | Mass | % dry | 72.22 | 27.78 | 100 |
| *b* | Ash | % dry | 45.0 | 83.4 | 55.6 |
| *c* | Energy Content (Kcal/kg) | dry | 4282 | 1291 | 3451 |
| | Energy loss in Reject | %Kcal/kg | | 10.4 | |
| *d* | Organic Matter Distribution | % | 94.1 | 5.9 | 100 |
| *e* | Mineral Matter Distribution | % | 58.4 | 41.6 | 100 |
| *f* | Baseline Efficiency | % | 33.54 | 33.54 | 33.54 |
| *g* | Moisture Correction | % | 98.8 | 98.8 | 98.8 |
| *h* | Boiler Correction | % | 93.31 | 73.84 | 89.18 |
| *i* | Auxillary Correction | % | 96.53 | 85.40 | 94.20 |
| *j* | Steam Correction | % | 98.26 | 94.44 | 97.40 |
| *k* | Total Correction | % | 0.87 | 0.59 | 0.81 |
| *l* | Net Efficiency | % | 29.32 | 19.73 | 27.11 |
| *m* | Gross Energy | Kcal/kg | 1155 | 234 | 861 |
| *n* | Gross Energy | Kwh/ton | 1217 | 247 | 907 |
| *o* | Gross Value | $/ton | 16.56 | 3.36 | 12.34 |
| *p* | -Freight Cost | $/ton | 10.08 | 10.08 | 10.08 |
| *q* | -Evaporation Cost | $/ton | 0.2 | 0.2 | 0.2 |
| *r* | -Ash cost | $/ton | 0.41 | 0.77 | 0.51 |
| *s* | -Miscellaneous | $/ton | 0.5 | 0.5 | 0.5 |
| *t* | NET Value | $/ton | 5.36 | (8.19) | 1.05 |
| | EMISSION ANALYSIS | | | | |
| *u* | Gross Energy | Kwh/lb | 0.61 | 0.12 | 0.45 |
| *v* | Emission/pound of coal | lb-$CO_2$/lb | 1.39 | 0.23 | 1.07 |
| *w* | Emission/unit energy | lb-$CO_2$/Kwh | 2.29 | 1.84 | 2.35 |
| *x* | Net $CO_2$ reduction | %lb-$CO_2$/Kwh | | 2.92 | |

**Figure 3.** An example of feasibility analysis for one of the pilot test at Kargali Washaery

As shown in the example, the true worth calculated on an equivalent energy content basis, the ROM feed is worth US $1.05/ton, while after deshaling the value of material increased to US $5.36/ton with organic matter recovery 94.1 % and a minimal energy loss in reject stream

10.4 % relative to feed. On the other side the reject stream after deshaling carry 83.4 % ash that values (negative) -US $8.19/ton. More notably, this stream is approximately 28 % of the ROM coal. Similarly, the power plant utilization efficiency improved from 27.11 % (for ROM) to 29.32 % (for deshaled product), shown in row 'I'. From these results, it is evident that the deshaling of ROM coal will be advantageous for the Indian coal industry.

Table 2 indicates economic gains and efficiency improvements using the feasibility model developed under the specified economic parameters for the pilot-scale unit. The analysis shows for the best operating conditions that include Test 13, Test 12, and Test 1 for Aryan, Bhushan and Kargali, respectively. As discussed above, the analysis showed net value increment (US $/ton) of 2.39, 1.81, and 4.31 for the de-shaled coal relative to ROM coal for the three test sites. The proposed dry deshaling method demonstrated significant economic gain especially for the case of Kargali coal samples.

**Table 2.** Economic, power plant efficiency, and emission analysis for the three test sites

| Item | Aryan energy | | Bhushan complex | | Kargali washery | |
|---|---|---|---|---|---|---|
| | ROM | De-shaled coal | ROM | De-shaled coal | ROM | De-shaled coal |
| Economic analysis (US $/ton) | 4.76 | 6.65 | 1.70 | 3.51 | 1.05 | 5.36 |
| Net value increment | | **2.39** | | **1.81** | | **4.31** |
| Power plant efficiency analysis (%) | 28.81 | 29.88 | 27.49 | 28.44 | 27.11 | 29.32 |
| Net efficiency improvement | | **1.07** | | **0.96** | | **2.21** |
| lb-$CO_2$/kwh % | 2.30 | 2.27 | 2.34 | 2.32 | 2.35 | 2.29 |
| lb-$CO_2$/kwh percent-reduction | | **1.57** | | **1.24** | | **2.92** |

Net value increment is calulated by substracting the ($/ton) value obtained from the analysis for de-shaled and ROM coal for the respective test sites

Similarly, Net efficiency improvement is calulated by substracting the power plant efficiency percentage obtained from analysis for the de-shaled and ROM coal for the respective test sites

lb-$CO_2$/kwh percent-reduction is a percentage-wise calculation using the following method: (ROM−De-shaled)*100/ROM

Net value increment is calulated by substracting the ($/ton) value obtained from the analysis for de-shaled and ROM coal for the respective test sites

Similarly, Net efficiency improvement is calulated by substracting the power plant efficiency percentage obtained from analysis for the de-shaled and ROM coal for the respective test sites

lb-$CO_2$/kwh percent-reduction is a percentage-wise calculation using the following method: (ROM–De-shaled)*100/ROM

These economic gains can be very significant when evaluated on yearly basis. For example, every 500-tph de-shaled coal can provide an additional gross revenue of $13 million annually (4.31 $/ton × 500 ton/h × 6000 h/year) for the coals like Kargali coal under specified economic parameters. Previous research showed an operating cost of a 500 tph unit approximately 0.5 $/ton ($1.5 million/year), and an initial capital cost ~3200 $/tph, *i.e.*, $1.6 million/year (Honaker 2007). As such, the payback period would be less than a year. Furthermore, due to inherently dry process, additional saving can be expected due to the reduction in usage of process water. It should be noted that the economic analysis would be a coal operation specific and need a detail market study for the operation to identify actual profit. The sole purpose here for these analyses to demonstrate benefits of deshaling qualitatively.

Likewise, power plant efficiency analysis suggested an improvement of 1.07, 0.96, and 2.21 percentage-points for Aryan, Bhushan, and Kargali coal samples, respectively, relative to their ROM coal. In addition, the proposed deshaling system offers direct cost reductions to power plants, since the removal of rock at mine face reduces grinding demands, improves boiler efficiency, increases carbon burn-out, reduces slagging/fouling problems, decreases erosion rates and lowers bottom ash/fly ash loads. More importantly, the calculated improvement in profitability resulting from the enhanced fuel quality provides an important financial incentive for Indian companies to implement this clean coal technology for other important purposes such as environmental control. Other control measures for reducing downstream emissions of $CO_2$, particulates and air toxics do not offer a similar financial return and require substantial capital investments that may be difficult to adopt given India's rapidly growing demands for energy infrastructure and other public services.

## 3.2 Emission analysis

For emission analyses, the specific energy generated (kwh/pound) from any given product can be estimated by using the net heat content of coal obtained from the feasibility studies (see Fig. 3 row '*u*'). Likewise, amount of $CO_2$ generated per unit of coal can be estimated thermodynamically (row '*v*'). The carbon content of the coal (assuming $C = 75\,\%$ of dry mineral-matter free basis) and the elemental ratio of $CO_2$ produced per unit of carbon (3.67), which is the theoretical limit used in the preliminary calculations. Based on these expressions, the amount of $CO_2$ emitted per unit of useful heat generated was determined by dividing unit of $CO_2$ emitted per unit of coal with specific useful energy generated (row '*w*'), followed by net emission reductions (row '*x*').

Table 2 provides summary of net reductions in emissions per unit of specific energy for the best results obtained from the three test sites. From the analysis, percent reduction in $CO_2$ emission can be achieved in the range of 1.2 %–3.0 % relative to the ROM coal, if deshaled coal from this technology is utilized for power generation. However, from a longer term perspective, the use of dry deshaling technologies to increase the availability of higher-quality coal feedstocks for downstream beneficiation process can serve as the first logical step in helping India to implement state-of-the-art clean coal technology processes to reduce total $CO_2$ emissions.

## 4 CONCLUSIONS

The pilot-scale testing of the inclined deck dry coal separator demonstrated the superior capability for deshaling and rejecting undesirable mineral matters from the Indian coarse (i.e. ≥6 mm) coal feedstocks currently utilized for thermal power generation. The testing was conducted at the three different coal preparation facilities in India, and showed successful recovery of the organic matter higher than 90 % while rejecting significant amounts of incombustible high-ash rock (i.e., 1 ton of waste rock rejected per 4–5 tons of higher-value clean coal recovered). The feasibility analysis indicated that the technology can provide significant economic gains as well as can improve the overall power plant efficiency by 1 %–3 %. Furthermore, emission rates of $CO_2$ from the power plants can also be reduced 1 %–3 % by implementing this dry coal separation approach.

## ACKNOWLEDGMENTS

Author expressed his gratitude to Eriez Manufacturing and Eriez India for the test equipment, field resources, and manpower that made much of this work possible. The author would like to acknowledge the tremendous support, field assistance and cooperation extended by the management and staff at Aryan Energy, Bhushan Steel & Power, and Central Coal Ltd., Kargali Washery. Finally, author is very thankful for the guidance and enormous support provided by Dr. Gerald Luttrell (E. Morgan Massey Professor, Virginia Tech), Mr. Robert Bratton (Consultant, Bratton Engineering, LLC), Dr. Tathagata Ghosh (Assistant Professor, University of Alaska Fairbanks), and Dr. Rick Honaker (Professor and Chair, University of Kentucky).

## FUNDING

The author would like to acknowledge the financial support for this work provided by the United States Department of State under the Asia–Pacific Partnership of Clean Development and Climate. Project number S-L-MAQM-07-GR-341.

## REFERENCES

1. Bhatt SM (2006) Effect of ash in coal on the performance of coal fired thermal power plants. Part I: primary energy Effects. Energy Sources 28(1):25–41

2. Biswal SK, Sahu AK, Parida A, Reddy PSR, Misra VN (2002) Prospects of dry beneficiation of Indian high ash non-coking coal: a review. J Mines Metals 51:53–57

3. Couch G (2002) Coal upgrading to reduce $CO_2$ emissions. International Energy Agency, CCC/67, Paris, France, pp 1–72

4. Dwari RK, Rao HK (2007) Dry beneficiation of coal —a review. Miner Process Extr Metall Rev 28(3):177–234

5. Dwari RK, Rao HK (2008) Non-coking coal preparation by novel tirbo-electrostatic method. Fuel 87(17):3562–3571

6. Gouri CT, Chattopadhyay US, Kabiraj SK, Haldar DD (2007) Dry beneficiation of non-coking coal with the Allair Jig—a case study. In: Proceedings of Advanced Gravity Separation, National Metallurgical Laboratory, Jameshedpur, India, pp 167

7. Gupta N (2011) Dry deshaling of thermal coals in India. Master Thesis, Virginia Polytechnic Institute and State University

8. Gupta N, Bratton R, Luttrell G, Ghosh T, Honaker R (2012) Application of air-table technology for cleaning Indian coals. In: Young C, Luttrell G (eds) Separation technologies for minerals, coals, and earth resources. Society for Mining, Metallurgy & Exploration, Englewood, CO, pp 199–210

9. Gupta N, Bratton R, Luttrell G, Ghosh T, Honaker R (2014) Upgrading Indian thermal coals using air-table dry deshaling. In: 31st International Pittsburgh Coal Conference, Oct. 2014, Pittsburgh, PA

10. Gupta N, Luttrell GH, Bratton RC (2016) Statistical optimization of dry deshaling for Indian coals using response surface methodology. Int J Coal Prep Util 36(3):125–138

11. Honaker RQ (2007) Coarse dry coal cleaning. Workshop on coal beneficiation and utilization of rejects: initiatives, policies and best practices, Ranchi, India, pp 22–24

12. Honaker RQ, Luttrell GH (2007) Development of an advanced deshaling technology to improve the energy efficiency of coal handling, processing, and utilization operations. University of Kentucky, Lexington

13. Honaker RQ, Luttrell GH, Lineberry GT (2006) Improved mining economics using near-face deshaling. Miner Metall Process 23(2):73–79

14. Honaker RQ, Saracoglu M, Thompson E, Bratton R, Luttrell GH, Richardson V (2008) Upgrading coal using a pneumatic density-based separator. Coal Prep Int J 28(1):51–67

15. IPCC (2013) The physical science basis, contribution of working group I to the fifth assessment report of the intergovernmental panel on climate change. Cambridge University Press, Cambridge

16. ISR (2005) Iron and steel review. 49(7), Hari Machines Limited, December 2005

17. MOC (2015) Ministry of Coal, Government of India. Accessed Sept 2015

18. MOP (2015) Ministry of Power, Government of India. Accessed Sept 2015

19. Sahu AK, Biswal SK, Parida A, Reddy PSR, Misra VN (2005) A study of dynamic stability of medium in air dense medium fluidised bed separator. India 56(1):103–107

20. Tiwari V, Wahr J, Swenson S (2009) Dwindling groundwater resources in northern India, from satellite gravity observations. Geophys Res Lett 36:L18401

21. Wills BA, Napier-Munn T (2006) Mineral processing technology, 7th edn. Elsevier, Amsterdam, p 65

22. World Bank (2015) Website accessed in September 2015 (comparative data for the year 2011)

23. Zamuda C, Sharpe M (2007) A case of enhanced use of clean coal in india: an essential step towards energy security and environmental protection. In: Workshop on Coal Beneficiation and Utilization of Rejects, pp 22–24

24. Zhao Y, Yang X, Luo Z, Duan C, Song S (2014) Progress in developments of dry coal beneficiation. Int J Coal Sci Technol 1(1):103–112

# Chapter 2

## Comparison of Pore Characteristics in the Coal and Shale Reservoirs of Taiyuan Formation, Qinshui Basin, China

**Yu Liu[1,2]•Yanming Zhu[1,2]**

[1] *Key Laboratory of Coal Methane Resource & Reservoir Formation Process of Ministry of Education, China University of Mining & Technology, Xuzhou 221008, Jiangsu, China*

[2] *School of Resource and Earth Science, China University of Mining & Technology, Xuzhou 221116, Jiangsu, China*

## ABSTRACT

Coal and shale are both unconventional gas reservoirs. Comparison of pore characteristics in shale and coal would help understand organic pore structure in shale and investigate co-exploration of shale gas and coalbed methane in coal bearing strata. In this study, five shale samples and three coal samples of Taiyuan Formation were collected from Qinshui Basin, China. High pressure mercury injection, scanning electronic microscopy, and fractal theory have been used to compare pore characteristics in shale and coal. The results show that pore volumes in coal are much larger than that in shale, especially pores 3–100 nm. In coal, there are many semi-closed pores in micro pores (<10 nm) and transition pores (10–100 nm). On the contrary, micro pores and transition pores are mainly with open pores in shale. The fractal curves show that pores larger than 65 nm in coal and shale reservoir both have obvious self-similarity and the fractal dimension values in shale and coal are similar. But the fractal characteristics of pores smaller than 65 nm in shale reservoir are quite different from that in coal.

**KEYWORDS:** Comparisoncoalshaleorganic Poresfractal

## 1 INTRODUCTION

Shale and coal seam are both unconventional gas reservoir for shale gas and coal bed methane. Shale gas and coalbed methane have the similar storage mechanism, and they are both mainly composed of free gas and adsorbed gas (Crosdale et al. 1998; Curtis 2002; Pan and Connell 2015; Pan and Wood 2015). Shale gas and coalbed methane are storied in pores, free gas staying in the center of pores and adsorbed gas adsorbed on the surface of pores (Zhang et al. 2012; Pan and Connell 2015). Pores in coal are all organic pores, while pores in shale having different types such as organic pores, interparticle pores and intraparticle pores (Loucks et al. 2012). Among them, the organic pores have a great significance in storage of shale gas (Loucks et al. 2009). Also the methane adsorption of shale is controlled by the organic matter (Zhang et al. 2012). Pore system in shale is more complex than that of coal because shales have both organic pores and inorganic pores (Loucks et al. 2012). For the heterogeneity of shale, it is difficult to study the organic pores and inorganic pores respectively. By comparing pores in shale and coal, it would help understand the organic pores characteristics in shale. Also in the marine- terrestrial sedimentary environment, organic shale is usually adjacent to the coal in the vertical. In view of this, many scholars put forward the idea of co-exploration of shale gas and coalbed methane in coal bearing strata (Zou et al. 2010; Qin et al. 2014). It is necessary to study the differences and similarity of pores in shale and coal by comparing them.

Benefit from exploration and development of coalbed methane and shale gas, studies on the pores in shale and coal have attracted much attention. In recent years, there have been many achievements on pores of coal and shale reservoir and the pore size distribution, pore types, fractal characteristics have been understood (Yao and Liu 2012; Xue et al. 2012; Clarkson et al. 2013; Zhang et al. 2014; Wang et al. 2014; Dong et al. 2015; Okolo et al. 2015; Pan et al. 2015; Sakhaee-Pour and Bryant 2015; Xiong et al. 2015). Coal and shale both have a wide pore size range, from nm scale to μm scale (Wang et al. 2014; Li et al. 2015). High pressure mercury injection experiment is one of the most powerful methods to characterize the pore structure in both shale and coal (Okolo et al. 2015). Also, scanning electron microscope (SEM) experiment makes the pores in shale and coal visual and brings great help for studies of pores.

Recently, fractal is commonly used to investigate the pore structure and surface irregularities of shale and coal (Mahamud and Novo 2008; ; Zhang et al. 2011; Pan et al. 2016; Li et al. 2015; Sun et al. 2015; Yang et al. 2014). Fractal theory is primary used to study nonlinear and irregular characteristics of shapes (Xie 1996; Pan et al. 2016). Mercury porosimetry and gas adsorption methods are the common methods to investigate the fractal feathers of pore structure of coal and shale (Bu et al. 2015; Yang et al. 2014). The investigation results of fractal methods on coal and shale showed that fractal theory is an effective tool to study pores structure in shale and coal (Bu et al. 2015; Pan et al. 2016), and the methane adsorption capacity of shale also has relations with the fractal dimension (Pan et al. 2016).

Few previous studies were focused on the comparison of pores in shale and coal reservoir, but it would help understand the organic pores in shale and be necessary in the study of co-exploration of shale gas and coalbed methane in coal bearing strata. Besides, pore structure of marine- terrestrial shale has not been understood clearly. In this study, we collected coal and shale samples from the coal bearing strata in Qinshui Basin, China. High pressure mercury and fractal analysis have been used to compare the pore characteristic in coal and shale reservoir.

# 2 SAMPLES AND EXPERIMENTAL

## 2.1 Samples

Five shale samples and three coal samples were chosen from Qinshui Basin in China in the study (Fig. 1). Qinshui Basin is located in the southeast of Shanxi Province and it is the most successful area for coalbed methane exploitation in China (Fig. 1). Qinshui Basin is a synclinore, in which fault is not developed.

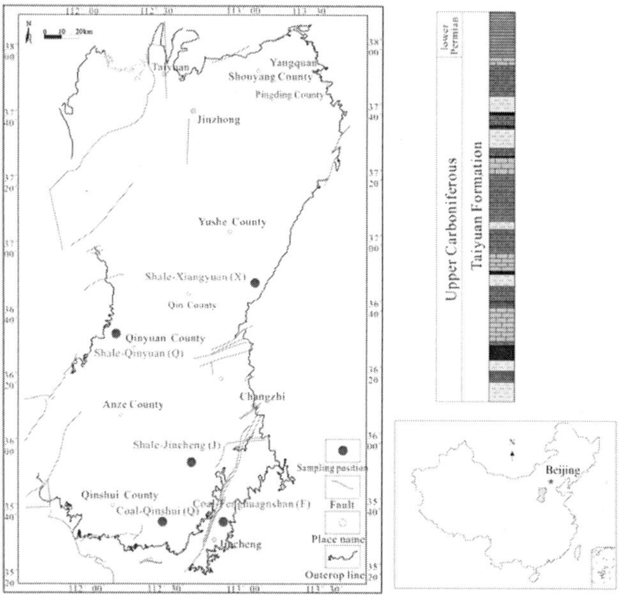


**Figure 1.** The sampling position and the typical stratum histogram of Taiyuan Formation in Qinshui

The Taiyuan Formation coal and shale in Qinshui Basin are both formed in the marine-continental transitional environment and they have the same type of kerogen (III). But there is a big difference between coal and shale in the composition of the material content. Clay content is the highest in shale, with average of 52.4 %, followed by the quartz content. TOC of the shale is mainly within 1.5 %–5.0 %. In coal, Organic matter is main component, and the content is above 90 %.

All of the shale samples are core samples and all of the coal samples were collected on the mining working face. The sample information is shown in Table 1. The shale and coal samples are all in stage of maturity to over-maturity.

**Table 1.** Sample information

| Sample number | Lithology | TOC (%) | *Ro* | Kerogen type | Clay content (%) | Quartz content (%) |
|---|---|---|---|---|---|---|
| J-1 | Shale | 3.44 | 2.92 | III | 51 | 29 |
| Q-1 | Shale | 1.72 | – | III | 58 | 30 |
| Q-2 | Shale | 2.12 | 1.72 | III | 43 | 30 |
| X-1 | Shale | 3.84 | 1.91 | III | 47 | 28 |
| X-2 | Shale | 2.58 | – | III | – | – |
| F-1 | Coal | – | 3.3 (Liu 2007) | III | – | – |
| F-2 | Coal | – | 3.3 (Liu 2007) | III | – | – |
| Q-1 | Coal | – | 2.4–2.8 | III | – | – |

*Note* The reflectance of vitrinite in coal sample Q-1 is from region data

## 2.2 Experiment

### 2.2.1 High pressure mercury injection experiment

Because of the nonwetting between mercury and coal, pressure is needed to make mercury into pores. The higher the pressure is, the smaller of the pore size which the mercury could enter will be. So we can measure the amount of mercury of a certain pressure to get the volume of corresponding pore size. The model of high pressure mercury injection experimental instrument is Auto Pore IV 9500 V1.09.

As the upper limit of pressure value is 6000 psi, the lower limit value of the measured pore size is 3.0 nm. The experimental samples are all the bulk samples, about 0.1–1 $cm^3$ in size. The quality of coal samples is 1.5–2.5 g, that of shale samples being 3–4 g. Samples need to be dried at 70 °C for more than 5 h before the experiment.

### 2.2.2 Scanning electron microscope (SEM)

Coal and shale samples for scanning electron microscope experiment are bulk samples. The SEM experiment was performed in Chinese University of Petroleum (Beijing). The experimental instrument is FEI Quanta 200. Some of the shale samples are polished by focused ion beam (FIB-SEM), while coal and some shale samples are not. Samples have been prepared by gold spray treatment before the experiment. When observing under scanning electron microscope, energy dispersive spectrometer (EDS) has been combined to judge the pore types.

### 2.2.3 Pore classification

In this study, we used the classification classified by Fu (2001): micropores <10 nm, transition pore being 10–100 nm, mesopores being 100–1000 nm, macropores are larger than 1000 nm.

## 3 RESULTS AND DISCUSSION

The high pressure mercury injection experimental results show that the porosity of coal samples in Qinshui Basin is 5.76 % on average, significantly larger than that of shale samples (1.65 %) (Table 2). Coal reservoir and shale reservoir of Taiyuan Formation in Qinshui Basin are both mainly with pores smaller than 100 nm (Fig. 2). Pores whose throat is 1000–10,000 nm are not developed in shale and coal reservoir. Mesopores (100–1000 nm) is well developed in coal reservoirs, but these pores are not developed in shale. Figure 2 also shows that in the stage of 3–100 nm, the number of pores in coal is significantly larger than that in shale. That is because organic pores are mainly 3–100 nm, and organic matter content in coal is much larger than that in shale. The increase of organic matter in coal brings much more organic pores, and these pores make the porosity of coal samples much larger than that of shale.

**Table 2**. Data of high pressure mercury injection of shale and coal

| Sample number | Lithology | Porosity (%) | Volume (mL/g) | Specific area ($m^2$) | Median pore diameter (volume) (nm) | Average pore diameter (nm) |
|---|---|---|---|---|---|---|
| F-1 | Coal | 4.66 | 0.04 | 19.67 | 7.5 | 7.2 |
| F-2 | Coal | 6.84 | 0.06 | 27.29 | 9.2 | 8.1 |
| Q-1 | Coal | 5.79 | 0.05 | 23.09 | 10 | 8.4 |
| J-1 | Shale | 1.88 | 0.01 | 2.38 | 191.1 | 15 |
| Q-1 | Shale | 1.74 | 0.01 | 3.2 | 13.6 | 10.2 |
| Q-2 | Shale | 1.17 | 0.01 | 1.55 | 26.2 | 13.7 |
| X-1 | Shale | 2.41 | 0.01 | 6.03 | 7.9 | 7.7 |
| X-2 | Shale | 1.4 | 0.01 | 2.34 | 16.4 | 10.9 |

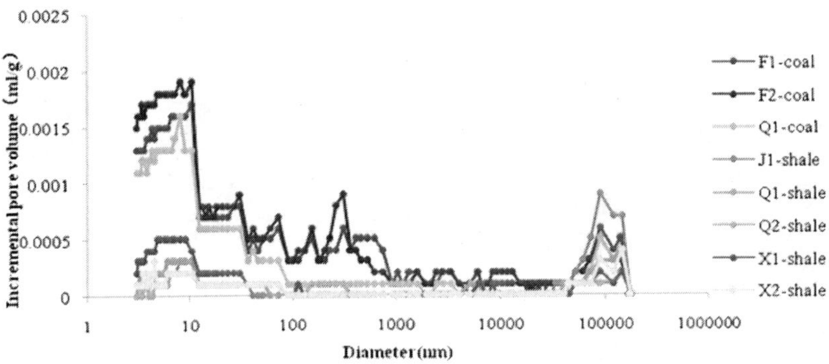


**Figure 2**. Pore structure distribution of coal and shale in Qinshui Basin

It should be noted that when testing the nanometer pores, the test pressure would become larger than 30 MPa and the high pressure mercury injection experiment would overestimate the pore volume for the compressibility in coal and shale. Okolo et al. (2015) has studied effects of coal compressibility on the pore structure distribution from mercury injection experiment. It was showed that the pore volume would be overestimated by about 20 %–25 %. Also, the pores could be damaged as the test pressure becomes larger than 30 MPa. The effects of damage in pore and compressibility are similar to coal and shale samples. Considering these factors, it is also clear that the pore volume of coal samples is much larger than that of shale samples, especially for pores smaller than 100 nm.

Coal and shale have great differences in genetical types of pores. As the coal is the organ-enriched reservoir, absolute advantage of organic matter content makes that the pores in coal are mainly with organic pores. In the shale, clay, quartz and other minerals content accounted for the major. In the shale reservoir, pores larger than 100 nm are mainly mineral intergranular pores, solution pores and clay pores, and some organic pores (Fig. 3a, b, d).

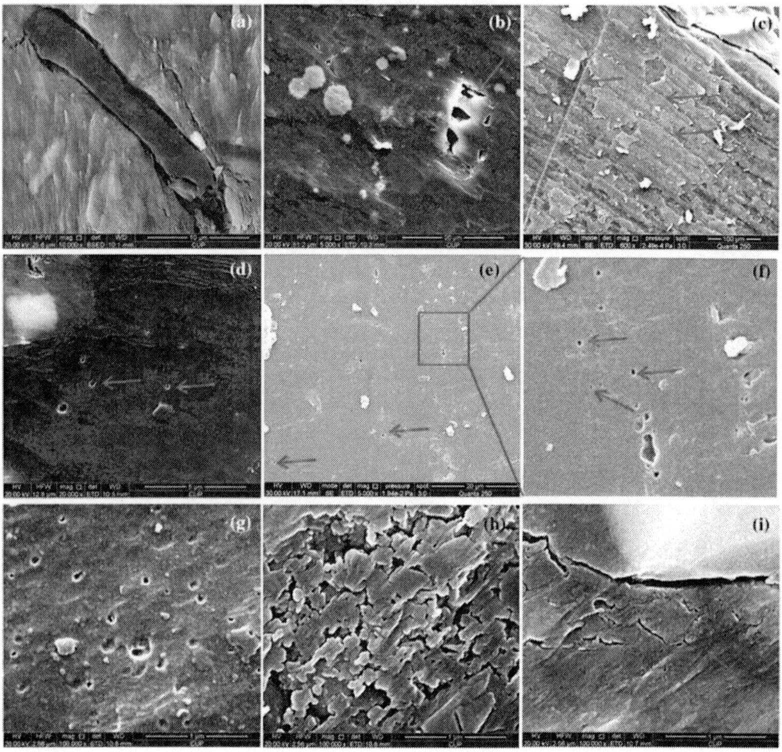


**Figure 3.** Pore of shale reservoir in the Taiyuan Formation from Qinshui Basin. *a* shale, FIB-SEM, intergranular pores. *b* shale, FIB-SEM, mineral dissolution pores *c* coal, SEM, cleats are well developed in the coal reservoir. *d* shale, FIB-SEM, organic pores. *e* coal, SEM, organic gas pores, familiar to organic pores in shale showed on d. *f* coal, SEM, organic gas pores, magnified part of c. *g* shale, FIB-SEM, organic pores, familiar to organic gas pores in coal reservoir showed on f. *h* shale, FIB-SEM, clay pores. *i* shale, FIB-SEM, clay pores

The organic matter type and maturity in coal and shale of Taiyuan Formation in Qinshui Basin is similar. Large numbers of organic gas pores (mainly transition pores and micropores) exit in both the coal reservoir and shale reservoir. Those gas pores in coal and shale are very similar in pore size and shape (Fig. 3d, e, f, g). For the same quality of coal and shale, the content of organic matter in the coal reservoir is 20–30 times that in shale reservoir. Given the similarity of the gas pores in coal reservoir and shale reservoir, the porosity of the 3–100 nm organic pores in coal reservoir should be 20–30 times of organic pores in shale reservoir. But in fact, the total volume and surface area of 3–100 nm pores in coal reservoir is about 10 times of that in shale, and this reflects that there are certain amounts of inorganic pores in shale. By calculating, the organic pores accounted for about 30 %–50 % of total 3–100 nm pores. Also we can observe many clay pores which are smaller than 100 nm by using SEM (Fig. 3h, i).

### 3.1 Mercury injection curves and mercury withdrawal curves

Mercury intrusion curves and mercury evacuation curves are often used in previous studies to reflect the pore shape of the pores in coal and shale reservoir (Chen et al. 2013; Qin 1994). Figures 4 and 5 show the mercury injection curves and mercury withdrawal curves of the coal and shale samples.

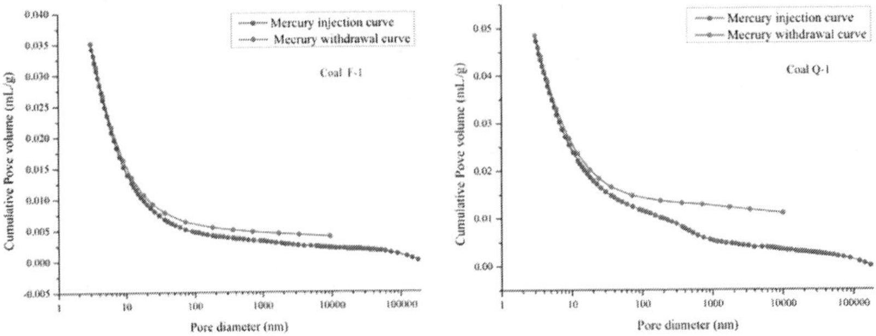


**Figure 4.** Mercury injection curve and mercury withdrawal curve of coal samples

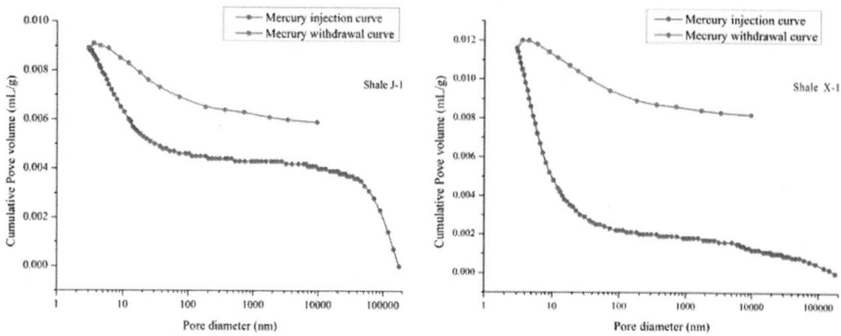


**Figure 5.** Mercury injection curve and mercury withdrawal curve of shale samples

The pore morphology would control the mercury injection- mercury withdrawal curves of the pores (Fig. 6). The differences in pore morphology of coal and shale samples would make the mercury injection-mercury withdrawal curves different. From Figs. 4 and 5, we can find that the mercury injection curves and mercury withdrawal curves of coal are quite different from that of shale, especially in the micro and transition pores stage. In micro and transition pores stage, mercury injection curve almost overlap mercury withdrawal curve in the coal reservoir, and this shows that there are a lot of semi-closed pres in the micro and transition pores (Qin 1994) (Fig. 6). But in the shale reservoir, the mercury injection curves are far from the mercury withdrawal curves (Fig. 5). At the beginning, the mercury injection curves show convex, and then have a sharp decline in 10 nm or so. It reflects that the pores at these stages in shale reservoir are mainly open pores, and also have many "flask" pores around 10 nm (Qin 1994). At the stage of larger than 100 nm, the mercury withdrawal curves of coal reservoir and shale reservoir all have certain distance from the mercury injection curves.

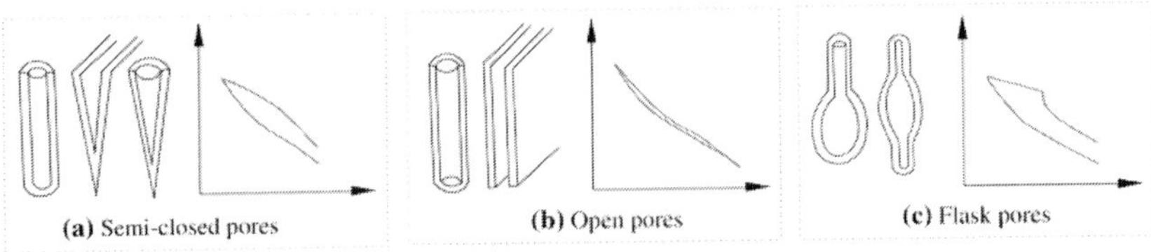


**Figure 6.** Pore morphology and mercury injection-mercury withdrawal curves (Qin 1994)

## 3.2 Fractal characteristics of pores in coal and shale reservoir

Coal and shale are both heterogeneous porous reservoir. For the heterogeneity of the shale and coal, fractal theory is introduced to study the pore characteristics of coal and shale (Mahamud and Novo 2008; Yang et al. 2014). In this study, high pressure mercury injection data of shale and coal samples was used to investigate the fractal features. The model used in the study is Menger sponge model, and the specific process of deduction and calculation are as follows (Xie 1996):

Assuming the length of initial element cube unit is *R*, divide the initial cube into $m^3$ small same cubes, then select a rule to remove part of the small cubes, and the number of cubes left would be *N1*; repeat the above operation on the smaller cubes. Then the side length of cubes decreases, and the cube number increases. After the operation for *K* times, the side length of small cube $r = R/mk$, and the number of $N_k = N_1k$, at the same time:

$$N_k = (R/r_k)^D = C_1/r_k^D = C_1\ r_k^{-D} \tag{1}$$

There into:

$$D = \lg N_1/\lg m,\ C_1 \text{ is a Constant} \tag{2}$$

$$V_k = N_k\ C_2\ r_k^3 = C_1\ C_2\ r_k^{3-D} \tag{3}$$

In the mercury injection experiment, pressure *P (R)* and aperture *r* satisfy the Washburn equation (Eq. 4):

$$P(r) = -2\sigma \cos \theta / r \tag{4}$$

In the formula, $\vartheta$ is mercury and solid medium contact angle (140 about) $\sigma$ is the surface tension of mercury (0.48 N/m).In the mercury injection experiment, mercury injection volume under certain pressure (*dVP (R))* is equal to the total pore volume (*V (k)*) at certain size:

$$dV(k) = dV_{P(r)} \tag{5}$$

$$dV_{P(r)}/dP(r) \propto [r/P(r)] \cdot r^{2-D} \tag{6}$$

$$D = 4 + lg(dV_p/dP(r))/lg(P(r)) \tag{7}$$

Through counting the mercury injection data for *dVp* (*R*)/*d* (*P* (*R*)) and *dP* (*R*), *lg*(*dVp* (*R*)/*P* (*R*)) – lg(*dP* (*R*)) diagram can be made, and the slope (K) of the curve can be calculated, and then the pore fractal dimension *D* would be obtained (Eq. (7)).

From Figs. 7 and 8 it could be found that the pore fractal curves of shale are similar to those of coal, and they all can be divided into two stages. But the boundary points of the two stages in shale and coal are different. The boundary points of coal reservoir fractal are in the interval of 54–85 nm, with an average of 65 nm, and that of shale reservoir pore fractal curve are smaller, 18–40 nm, with an average of 28 nm.

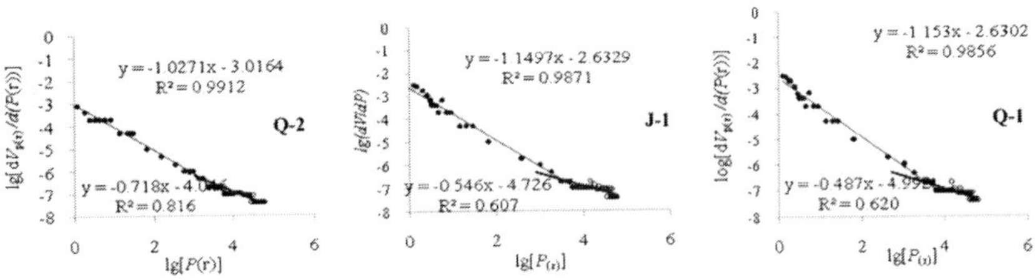


**Figure 7.** Pore fractal curve of shale reservoir

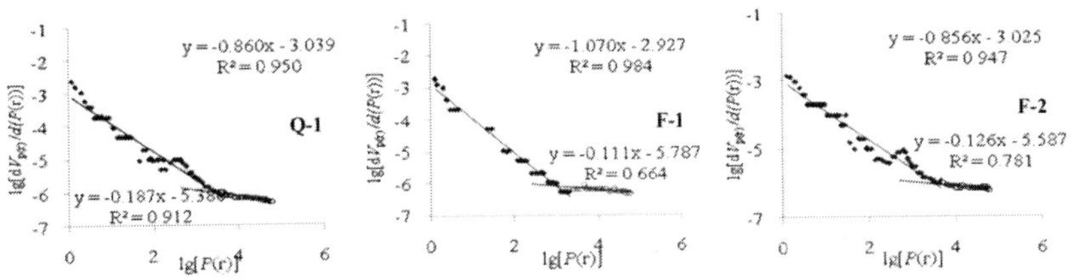


**Figure 8.** Pore fractal curve of coal reservoir

The pores larger than 65 nm in coal and shale reservoir both have obvious self-similarity: the degree of association between the pore fractal curves and the fitting line is larger than 0.9. The fractal dimension values of pores in coal are 2.93–3.14, and fractal dimension values of pores in shale are 2.99–3.12. The fractal dimension values of shale are similar to that of coal, and this reflects that surface roughness of pores larger than 65 nm in both coal and shale is very similar.

Based on the fractal principle, the calculated fractal dimension for this model can't be larger than 3, but some calculated results shows the fractal dimensions are larger than 3. The reasons of these phenomena have been discussed in previous studies (An et al. 2011; Wang and He 1996). On the one hand, as the pores measured by mercury injection experiment are all connected pores, closed pores in the reservoir cannot be measured. But the fractal model used in the calculation (Menger sponge model) contains not only connected pores, but also closed pores. Thus the fractal dimension calculated by using the mercury injection data has slight deviations compared to the practical dimension. On the other hand, in the mercury injection experiments, with decrease of test aperture, the mercury pressure increases. In this case, smaller pores would be overestimated for the compressibility of coal and shale samples. As fractal dimension value is very close to 3, slight deviations would make the fractal dimension calculated larger than 3.

The fractal characteristics of pores smaller than 65 nm in the coal reservoir are very different from that in shale reservoir. In the coal reservoirs, pores between 28 and 65 nm do not have self-similarity, but the self-similarity of shale pores in this stage is still obvious. This reflects that pores 28–65 nm in coal and shale reservoir have differences in shape. When the pore size is smaller than 28 nm, the fractal curves of shale are similar to that of coal samples. The reasons are that although the pores of the 28–65 nm in shale reservoir have a lot of organic pores, but inorganic pores still occupies a large part of the total pores. With the decrease in pore size, the proportion of organic pores increases continuously. When the pores size is smaller than 28 nm, pores in shale are mainly with organic pores, and the fractal curves of shale are similar to that of coal samples.

## 4 CONCLUSIONS

The porosity of coal samples from Qinshui Basin is 5.76 % on average, and it is significantly larger than that of shale samples (1.63 %). Pore structure distribution of coal and shale samples show that pores smaller than 100 nm in coal are far more than that in shale.

Mercury injection curves and mercury withdrawal curves of shale samples are quite different from that of coal samples. In micro and transition pores stage, mercury injection curve almost overlap mercury withdrawal curve in the coal reservoir. There are a lot of semi closed pores in micro pores and transition pores in coal, but in shale micro pores and transition pores are mainly with open pores.

Pores larger than 65 nm in coal and shale reservoir have similar fractal dimension values. But the fractal characteristics of pores smaller than 65 nm in coal are different from that in shale. That is because inorganic pores still occupy part of the pores smaller than 65 nm in shale. With the decrease in pore size, the proportion of organic pores increases in shale. When the pores size is smaller than 28 nm, pores in shale are mainly with organic pores, and the fractal curves of shale are similar to that of coal samples.

## ACKNOWLEDGMENTS

The authors thank the National Science Foundation of China (41472135), the Research and Innovation Project for College Graduates of Jiangsu Province (KYLX15_1396), the Scientific Research Foundation of the Key Laboratory of Coalbed Methane Resources and Reservoir Formation Process, Ministry of Education (China University of Mining and Technology) (No. 2015-04) for the support of the research.

## REFERENCES

1. An S, Sang S, Li Y, Wang G (2011) Study on pore fractal characteristics of high-rank coal reservoirs in Southern Qinshui Basin. Coal Geol China 23(2):17–21

2. Bu H, Ju Y, Tan J, Wang G, Li X (2015) Fractal characteristics of pores in non-marine shales from the Huainan coalfield, eastern China. J Nat Gas Sci Eng 24:166–177

3. Chen S, Xia X, Qin Y, Wang Y (2013) Classification of pore structures in shale gas reservoir at the Longmaxi Formation in the south of Sichuan Basin. J China Coal Soc 38(5):760–765

4. Clarkson CR et al (2013) Pore structure characterization of North American shale gas reservoirs using USANS/SANS, gas adsorption, and mercury intrusion. Fuel 103:606–616

5. Crosdale PJ, Beamish BB, Valix M (1998) Coalbed methane sorption related to coal composition. Int J Coal Geol 35:147–158

6. Curtis JB (2002) Fractured shale-gas systems. AAPG Bull 86(11):1921

7. Dong Q, Zhang H, Zhu Z (2015) Evolution of structure properties during Zhundong coal pyrolysis. Procedia Eng 102:4–13

8. Fu X (2001) Physical simulation of multiphase medium coal rock properties and numerical simulation. University of Mining and Technology, Xuzhou

9. Li W, Liu H, Song X (2015) Multifractal analysis of Hg pore size distributions of tectonically deformed coals. Int J Coal Geol 144–145:138–152

10. Liu F (2007) The characteristics of coal reserviors and evaluation of coalbed methane enrichment and high-productivity in QinShui Basin of ShanXi Province. Chengdu University of Technology, Chengdu

11. Loucks RG, Reed RM, Ruppel SC, Jarvie DM (2009) Morphology, genesis, and distribution of nanometer-scale pores in siliceous mudstones of the Mississippian Barnett Shale. J Sediment Res 79(12):848–861

12. Loucks RG, Reed RM, Ruppel SC, Hammes U (2012) Spectrum of pore types and networks in mudrocks and a descriptive classification for matrix-related mudrock pores. AAPG Bull 96(6):1071–1098

13. Mahamud MM, Novo MF (2008) The use of fractal analysis in the textural characterization of coals. Fuel 87(2):222–231

14. Okolo GN, Everson RC, Neomagus HWJP, Roberts MJ, Sakurovs R (2015) Comparing the porosity and surface areas of coal as measured by gas adsorption, mercury intrusion and SAXS techniques. Fuel 141:293–304

15. Pan Z, Connell LD (2015) Reservoir simulation of free and adsorbed gas production from shale. J Nat Gas Sci Eng 22:359–370

16. Pan Z, Wood DA (2015) Coalbed methane (CBM) exploration, reservoir characterisation, production, and modelling: a collection of published research (2009–2015). J Nat Gas Sci Eng 26:1472–1484

17. Pan J, Zhu H, Hou Q, Wang H, Wang S (2015) Macromolecular and pore structures of Chinese tectonically deformed coal studied by atomic force microscopy. Fuel 139:94–101

18. Pan J et al (2016) Micro-pores and fractures of coals analysed by field emission scanning electron microscopy and fractal theory. Fuel 164:277–285

19. Qin Y (1994) Micropetrology and structural evolution of high-bank coals in P.R.China. China University of Mining and Technology press, Beijing

20. Qin Y, Liang J, Shen J, Liu Y, Wang C (2014) Gas logging shows and gas reservoir types in tight sandstones and shales from Southern Qinshui Basin. J China Coal Soc 39(8):1559–1565

21. Sakhaee-Pour A, Bryant SL (2015) Pore structure of shale. Fuel 143:467–475

22. Sun W, Feng Y, Jiang C, Chu W (2015) Fractal characterization and methane adsorption features of coal particles taken from shallow and deep coalmine layers. Fuel 155:7–13

23. Wang E, He X (1996) Fractal desription of coal seam porosity and crack system and its application. J Fuxin Min Inst (Nat Sci) 15(4):407–410

24. Wang Y, Zhu Y, Chen S, Li W (2014) Characteristics of the nanoscale pore structure in Northwestern Hunan Shale gas reservoirs using field emission scanning electron microscopy, high-pressure mercury intrusion, and gas adsorption. Energy Fuels 28(2):945–955

25. Xie H (1996) Introduction to fractal petrology. Science Press, Beijing

26. Xiong J, Liu X, Liang L (2015) Experimental study on the pore structure characteristics of the Upper Ordovician Wufeng Formation shale in the southwest portion of the Sichuan Basin, China. J Nat Gas Sci Eng 22:530–539

27. Xue G, Liu H, Li W (2012) Deformed coal types and pore characteristics in Hancheng coalmines in Eastern Weibei coalfields. Int J Min Sci Technol 22(5):681–686

28. Yang F, Ning Z, Liu H (2014) Fractal characteristics of shales from a shale gas reservoir in the Sichuan Basin, China. Fuel 115:378–384

29. Yao Y, Liu D (2012) Comparison of low-field NMR and mercury intrusion porosimetry in characterizing pore size distributions of coals. Fuel 95:152–158

30. Zhang S, Yang S, Cheng J, Zhang B, Lu C (2011) Study on relationships between coal fractal characteristics and coal and gas outburst. Procedia Eng 26:327–334

31. Zhang T, Ellis GS, Ruppel SC, Milliken K, Yang R (2012) Effect of organic-matter type and thermal maturity on methane adsorption in shale-gas systems. Org Geochem 47:120–131

32. Zhang S, Tang S, Tang D, Huang W, Pan Z (2014) Determining fractal dimensions of coal pores by FHH model: problems and effects. J Nat Gas Sci Eng 21:929–939

33. Zou C, Dong D, Wang S, Li J (2010) Geological characteristics and resource potential of shale gas in China. Pet Explor Dev 37(6):641–653

# Chapter 3

# Recovery of Metals and Other Beneficial Products from Coal Fly Ash: A Sustainable Approach for Fly Ash Management

**Prafulla Kumar Sahoo[1,2] • Kangjoo Kim[2] • M. A. Powell[3] • Sk Md Equeenuddin[4]**

[1] *Vale Institute of Technology, Baventura da Silva, Belem, Para 66055-090, Brazil*

[2] *Department of Environmental Engineering, Kunsan National University, Jeonbuk 573-701, Republic of Korea*

[3] *Prometeo, SDIC, Secretary of Higher Education, Science, Technology and Innovation (SENESCYT), Quito, Ecuador*

[4] *Department of Mining Engineering, National Institute of Technology, Rourkela, Orissa 769008, India*

## ABSTRACT

Increasing production and disposal of coal fly ash (CFA) is a matter of serious environment concern. However, CFA contains various beneficial metals and mineral matters whose demand is increasing in the industrialized world, while natural supplies are diminishing. Therefore, recovery of these potential resources from CFA can be an alternative way to save mineral resources, as well as to reduce the environmental burden of CFA disposal. There are numerous methods developed for the recovery of beneficial products from CFA. Based on the US patents and journal literatures, the present review describes the recovery status and technologies of major elements such as Al, Si, Fe and Ti, and trace elements such as V, Ga, Ge, Se, Li, Mo, U, Au, Ag, Pt groups and rare earth elements (REEs) and other beneficial products such as magnetic materials, cenospheres, and unburned carbon from CFA. It also highlights the recovery efficiency and drawbacks for their extraction, and suggests future research to develop satisfactory results in terms of selective recovery and purification.

**Keywords** : Coal fly ash Beneficial elements Recovery Hydrometallurgical methods Extraction efficiency

# 1 INTRODUCTION

Coal based thermal power plants are one of the largest sources of electricity, contributing around 39 % of the total electricity generated throughout the world (Fig. 1) (US EIA 2014). This is due the fact that coal is the most abundant fossil fuel, and it is comparatively much lower priced than other fuels such as oil and natural gas. However, growing use of coal in power plants is generating huge amounts "waste", known as coal fly ash (CFA), with China, the US and India being the largest producers (Heidrich et al. 2013) (Table 1). The current global annual production of CFA is approximately 750 million tonnes (Yao et al. 2015), and this is anticipated to increase in the near future. Thus, increases in the global production of CFA is a matter of serious environmental concern because of issues related not only to its disposal but also the threat to public health and potential damage to soil, agriculture and natural beauty. These problems are more critical in areas experiencing increasing land use competition. Fly ash can even reach the sub-soil and contaminate the groundwater with heavy metals (Prasad and Mondal 2009; Ramya et al. 2013). Thus, there is a pressing need of its recycling and utilization (Ilic et al. 2003; Yao et al. 2015). More efforts have been undertaken on concrete production, paving roads, mine fillings, building material production, land reclamation, soil stabilization, toxic element immobilization, and synthesis of polymers and agriculture etc. (Kruger 1997; Basu et al. 2009; Ahmaruzzaman 2010; Alam and Akhtar 2011; Loya and Rawani 2014). Taking this into account, on average only 25 % of CFA is utilized (Sommerville et al. 2013), wherein 38 % for US, 47 % for the EU and 15 % for India (Yao et al. 2014a, b; reference therein). The rest is considered as pollutant and disposed waste. Hence, there is an urgent need for developing of its novel applications (Blissett et al. 2014).

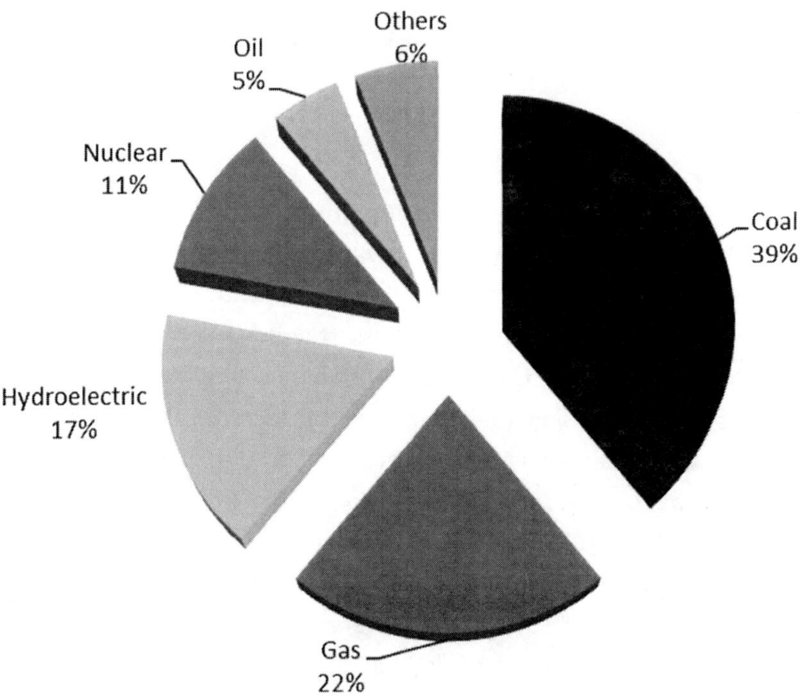


**Figure 1.** World electricity production from all energy in 2014 (US EIA 2014)

**Table 1.** Worldwide annual production of coal fly ash (Heidrich et al. 2013)

| Country | Ash production (Mt) |
|---|---|
| China | 385 |
| USA | 118 |
| India | 105 |
| Europe | 52.6 |
| Middle East and Africa | 32.6 |
| Russian Federation | 26.6 |
| Other Asia | 16.7 |
| Australia | 13.1 |
| Japan | 11.1 |
| Canada | 6.8 |

Currently, CFA is being explored as an alternative source of valuable metals, since it mainly composes of a variety of oxygenated inorganic compounds, such as $SiO_2$, $Al_2O_3$, $Fe_2O_3$ and $TiO_2$, with smaller amounts of trace elements (Lai-shi et al. 2011; Blissett and Rowson 2012; Seredin et al. 2013; Mayfield and Lewis 2013). With diminishing metal resources as well as ever-increasing their demand, recovery of these products from CFA would be highly worthwhile and a challenging topic for current international research (Cao et al. 2008; Mayfield and Lewis 2013; Tolhurst 2015). Nevertheless, recovery of metals from CFA can reduce environmental risks associated with disposal and leaching of metals to surface and subsurface waters. There have been several reviews based on journal articles that focused on the recovery of beneficial elements and products from CFA (Jha et al. 2008; Cao et al. 2008; Meawad et al. 2010; Blissett and Rowson 2012; Yao et al. 2014a, b; Franus et al. 2015). However, the present review focused on US Patents, as well as journal articles to present an overview of the recovery status and technologies of metals and other beneficial products from CFA using several metallurgical and bio-metallurgical methods with the aim of evaluating their extraction efficiency, impact and applicability.

## 2 PROPERTIES OF COAL FLY ASH

The physico-chemical properties of CFA are highly variable depending on coal type, combustion temperature, combustion technique, air/fuel ratio, and particle size of the feed coal (Dhadse et al. 2008). The major oxides for CFA from different coal types are presented in Table 2 (Meawad et al. 2010). The principal components of all ash samples are $SiO_2$, $Al_2O_3$, and $Fe_2O_3$, with varying amounts of carbon, reported as loss on ignition (LOI), CaO, and MgO. Higher concentrations of CaO and MgO, and lower concentrations of $SiO_2$, $Fe_2O_3$ and carbon are found in lignite and sub-bituminous CFA compared to bituminous CFA. According to the American Society of Testing Materials (ASTM C618 2000), two classes of CFA have been defined based on the coal usage. Fly ash produced from anthracite or bituminous coal is classified as class F, while class C fly ash is produced from lignite or sub-bituminous coal. This classification is based on differences in contents $SiO_2$, $Al_2O_3$, and $Fe_2O_3$, and LOI in the ash. In classes C and F, the $SiO_2 + Al_2O_3 + Fe_2O_3$ contents are >50 % and >70 % respectively, and LOI is <6 % and <12 %, respectively. In addition, class C fly ash corresponds to high-lime (10 % up to 40 %), while Class F fly ash has lime concentrations <10 % CaO. Coal fly ash is considerably rich in trace elements (Karayigit et al. 2001; Vassilev et al. 2005; Dhadse et al. 2008). The normal range of those elements is given in Table 3 (National Research Council 2006; USEPA 1999; Meawad et al. 2010). Elements in CFA are enriched mainly based on their organic/inorganic affinity and host mineralogy, followed by volatility (Xu et al. 2003; Vassilev et al. 2005). During fly ash gasification, the majority of mineral fractions hosting important trace elements is volatilized and then condensed on the surfaces of the fly ash (Clark and Sloss 1992; Querol et al. 1995). US Patent No. 0274886A1 (2007)

**Table 2.** Major oxide concentrations in fly ash from different coal types (%) (Meawad et al. 2010; references therein)

| Component | Bituminous | Sub-bituminous | Lignite |
|---|---|---|---|
| $SiO_2$ | 20–60 | 40–60 | 15–45 |
| $Al_2O_3$ | 5–35 | 20–30 | 10–25 |
| $FeO_3$ | 10–40 | 4–10 | 4–15 |
| CaO | 1.12 | 5–30 | 15–40 |
| MgO | 0–5 | 1–6 | 3–10 |
| $Na_2O$ | 0–4 | 0–2 | 0–6 |
| $K_2O$ | 0–3 | 0–4 | 0–4 |
| LOI | 0–15 | 0–3 | 0–5 |

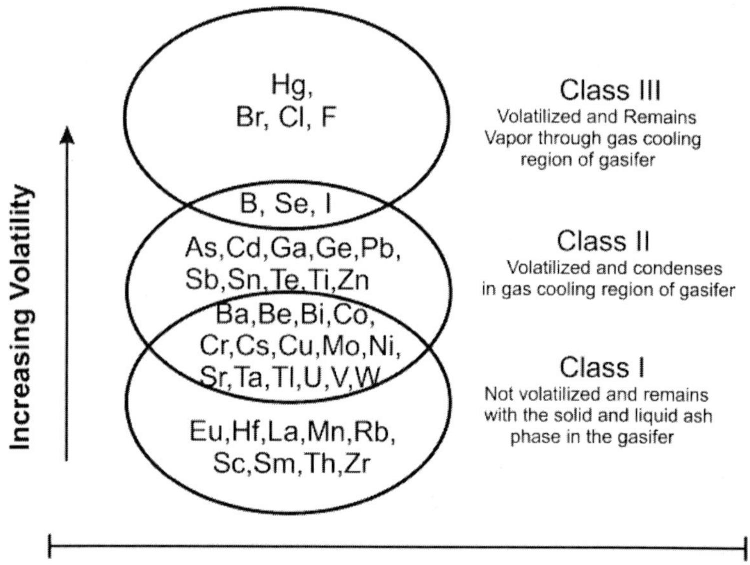


**Figure 2.** Relative volatility of trace elements in coal gasification systems (US Patent 0274886A1 2007)

**Table 3.** Trace element concentrations in fly ash (mg/kg) (National Research Council 2006; USEPA 1999; Meawad et al. 2010)

| Element | Median | Range |
|---|---|---|
| As | 43.4 | 0.0003–391 |
| Ba | 806.5 | 0.02–10,850 |
| Be | 5 | 0.2–2105 |
| B | 311 | 2.98–2050 |
| Cd | 3.4 | 0.01–79 |
| Cr | 136 | 3.6–437 |
| Co | 35.9 | 4.9–79 |
| Cu | 112 | 0.2–655 |
| F | 29 | 0.04–320 |
| Pb | 56.8 | 0.02–273 |
| Mn | 250 | 24.5–750 |
| Hg | 0.1 | 0.013–49.5 |
| Ni | 77.7 | 0.1–1270 |
| Se | 7.7 | 0.0003–46.5 |
| Sb | 4.6 | 0.2–205 |
| Ag | 3.2 | 0.01–49.5 |
| Sr | 775 | 30–3885 |
| Tl | 9 | 0.15–85 |
| V | 252 | 43.5–5015 |
| Zn | 148 | 0.28–2200 |

illustrates the volatility classes of elements during combustion (Fig. 2). This indicated that elements such as As, B, Bi, Cd, Ge, Ga, Mo, Pb, S, Se, Sb, Sn, Te, and Ti are classified into Class II category. These elements have tendency to more concentrate in the finer particles of the ashes during the condense-volatilization process (US Patent No. 4686031 1987; Gieré et al. 2003).

# 3 RECOVERY OF MAJOR ELEMENT

There are a number of methods have been developed for the recovery of major elements from CFA (Table 4), which are discussed below.

**Table 4.** Recovery of major and trace elements, their extraction methods and recovery efficiency from coal fly ash

| References | Element | Method | Chemical reagent/others | Recovery efficiency |
|---|---|---|---|---|
| Burnet et al. (1977) | Al | High-temperature chlorination | Carbon, chlorine | Not known |
| Mehrotra (1982a, b) | Al, Ti | High-temperature chlorination | Carbon, carbon monoxide, chlorine | 25 % Al, 25 % of Ti |
| US Pat. No. 4159310 (1979) | Al, Ti | High-temperature chlorination | Carbon monoxide, chlorine, $H_2SO_4$ | 70 %–80 % Al, 80 % Ti |
| US Pat. No. 4243640 (1981) | Al | Magnetic separation, acid leaching, calcination | $HNO_3$ | Up to 90 % Al |
| US Patent No. 4567026 (1986) | Al, Ti | Magnetic separation, leaching, calcination | $H_2SO_4$, NAOH, $CO_2$ | Not known |
| US Patent No. 4254088 (1981) | Al, Ti | Salt-soda sintering, leaching | NaCl, $Na_2CO_3$, $HNO_3/H_2SO_4$ | 90 % Al |
| Matjie et al. (2005) | Al, Ti | Calcination, leaching, solvent extraction | CaO, $H_2SO_4$ | 99.4 % $Al_2O_3$, 97 % $TiO_2$ |
| US Pat. No. 4539187 (1985) | Al, Ti | Acid leaching | Fluorosilicic acid, HF | Not known |
| Wu et al. (2012) | Al | Pressure acid-leaching | $H_2SO_4$ | 82.4 % Al |
| US Pat. No. 4130627 (1978) | Al, Fe | Alkaline leaching, electrolysis | NaOH, HCl/HF | 65 % Fe, 89 % Al |
| Padil and Sohn (1985) | Al | Lime-soda sintering | $CaCO_3$, $NaCO_3$ | 50 %–60 % alumina |
| Lu et al. (2003) | Al | Lime-soda sintering | Limestone, $NaHCO_3$ | 70 % alumina |
| Bai et al. (2010) | Al | Lime-soda sintering | Lime, $NaCO_3$, NaOH | 90 % alumina |
| US Pat No. 4397822A (1983) | | Lime-soda sintering | $CaCO_3$, $NaCO_3$, Sulfur, carbon | 90 % alumina |
| US Pat No. 0119426A1 (2010) | Si, Al | Leaching, calcination | NaOH, $CO_2$, CaO, $H_2SO_4$ | Not known |
| US Pat No. 7871583B2 (2011) | Si, Al | Leaching, calcination | NaOH, $CO_2$, CaO, $H_2SO_4$ | Not known |
| US Pat No. 5231018 (1993) | Al, Ti | Bioleaching | C and N source, GIN-1, growth medium | Not known |
| US Pat No. 5278069 (1994) | Al, Ti | Bioleaching, acid precipitation | *Thiobacillus thiooxidans* | Not known |
| US. Pat No. 4539187 (1985) | Si | Acid leaching | $SiF_4$, HF | Not known |
| US Pat. No 0287827A1 (2010) | Si | Acid leaching | HF, coal, S | Not known |
| US Pat. No. 0119426A1 (2010) | Si | Alkaline leaching, carbonation | NaOH, $H_2SO_4$, CaO, $CO_2$ | >98.5 % $SiO_2$ (purity) |
| US Pat. No. 7871583B2 (2011) | Si | Alkaline leaching, carbonation | NaOH, $H_2SO_4$, CaO, $CO_2$ | >98.5 % $SiO_2$ (purity) |
| US Pat. No 4814150 (1989) | V | Leaching, precipitation | NaOH, $H_2SO_4$, ammonium salt | 80 %–90 % |
| US Pat. No. 4966761 (1990) | V | Leaching, solvent extraction | NaOH, resin, HCl, NH4Cl, quaternary amine | Up to 93.4 % V |
| US Pat. No. 7498007B2 (2009) | V | Acidic leaching | Calcium hydroxide, quick lime, acid | >98 % |
| Xiang-yang et al. (2010) | V | Acid leaching | $H_2SO_4$ | 95 % |
| Vitolo et al. (2000) | V | Combustion, acid leaching | $H_2SO_4$, $NaClO_3$ | 83 % |
| US Pat. No. 6652819B2 (2003) | V | Combustion, acid leaching | $H_2SO_4$ | Not known |
| US Pat No. 4475993 (1984) | Ga, others | Alkali halite melt, electrolysis | Aluminium trichloride | 69 % of Ga |
| US Pat No. 4643110 (1987) | Ga, Ge | Sublimation | | Not known |
| US Pat No. 4678647 (1986) | Ga, Ge | Sublimation | | Up to 90 % Ga |
| US Pat No. 4757772 (1988) | Ge, Ga | Sublimation | | Not known |
| US Pat No. 4678647 (1986) | Ge, Ga | Sublimation | | Up to 90 % Ga |

| | | | | |
|---|---|---|---|---|
| Arroyo et al. (2014) | Ge, Ga | Acid, alkaline, complexant leaching | Oxalic acid, sulfuric acid | Up to 90 % Ge; up to 82 % Ga |
| Gutiérrez et al. (1997) | Ga | Acid leaching | HCl, Amberlite LA-2 | 83 % of Ga |
| Fang and Gesser (1996) | Ga | Acid leaching, flotation | HCl | Up to 80 % of Ga |
| Arroyo et al. (2009) | Ge | Water leaching, solvent extraction | Water, catechol, NaOH | 95 % Ge |
| Font et al. (2005) | Ge | Water leaching | | 86 % of Ge |
| Mok et al. (1984) | Mo | Acid digestion, solvent extraction | HCl, $HNO_3$, dithiocarbamate | Not known |
| Ogata et al. (2011) | Mo | Acid/alkaline leaching | HCl, NaOH, gibbsite | 90 % of Mo |
| US Pat. No. 3857920 (1974) | Li | Leaching | Water, $CO_2$ | 80 %–90 % of $Li_2CO_3$ |
| Wang et al. (2007) | Se | Leaching | $HNO_3$, NaOH | 55 %–69 % at pH 12 |
| Sparton Res Inc. (2007) | U | Acid leaching, ion-exchange | $H_2SO_4$, resin | Not known |
| Maslov et al. (2010) | U, Ra | Acid leaching | $HNO_3$, HF | U (99 %), Ra (97 %) |
| Us Pat No. 6827837B2 (2004) | Au, Pt group | Combustion, acid digestion, precipitation | $HNO_3$, HCl | Not known |
| US Pat No. 0056548A1 (2005) | Au, Pt group | Combustion, acid digestion, precipitation | $HNO_3$, HCl | Not known |
| US Pat No. 4649031 (1987) | REEs | Sieving, aqueous alkali/acid digestion | $Na_2CO_3$, HCl, $HNO_3$ | Not known |
| US Pat No. 0287653A1 (2013) | REEs | Acid leaching, Ion-exchange | $HNO_3$, tributyl phosphate, kerosene | Not known |
| US Pat No. 8968688B2 (2015) | REEs | Acid leaching, Ion-exchange | $HNO_3$, tributyl phosphate, kerosene | Not known |
| Narukawa et al. (2007) | Cr | Alkaline leaching | NaOH | Not known |
| Huang et al. (2011) | Pb, Zn | Alkaline leaching | NH4OH/NaOH | Not known |
| Okada et al. (2007) | Zn, Pb | Acid leaching | $CH_3COOH$ | 97 % Zn, 98 % Pb |
| Hong et al. (2000) | Cr, Cu, Pb, Zn | Leaching | Chelating agents (EDTA) | 20 %–50 % Cr, 60 %–95 % Cu<br>60 %–100 % Pb, 50 %–100 % Zn |
| US Pat. No. 5278069 (1994) | Co, others | Bioextraction | *Thiobacillus thiooxidans* | Not known |

## 3.1 Aluminum and titanium

Aluminum (Al) is an industrially important metal; it is light, corrosion resistant, and malleable and is mainly used in the aircraft industry. It is also used in the production of aluminum chemicals and in refractories (Matjie et al. 2005). Titanium (Ti) is used for various purposes in industry. In particular, it has been used in the aerospace industry. Titanium oxide is used as a pigment in the paint industry. On larger-scale, Ti and titanium alloy products are used for making automotive parts (Fontana et al. 2005). Coal and coal residues such as CFA are enriched with these elements (Halina et al. 2007; Seredin 2012; Dai et al. 2012a).

Several methods have been reported for recovering Al from CFA. Burnet et al. (1977) introduced a chlorinating process, where the fly ash is first separated into magnetic and non-magnetic fractions. The non-magnetic fraction of CFA was mixed with carbon and chlorinated in a fixed bed. The iron remaining in this fraction was removed by lowering the temperatures between 400 and 600 °C. The chlorinated aluminum fraction was recovered by increasing the temperature to 850–950 °C. Following Burnet et al. (1977), Mehrotra et al. (1982a,b) studied a high-temperature chlorinating process, and found that only 25 % of Al and 25 % of Ti in ash could be chlorinated at temperature above 900 °C. However, this process left about one-third of the iron with the nonmagnetic fraction. US Patent No. 4159310 (1979) gives a more thorough discussion of the chlorinating process for the recovery of Al and Ti. In this Patent the nonmagnetic fraction of ash was chlorinated in an oxidizing atmosphere and selectively separated Fe from other metals. The remaining residue was then chlorinated in a reducing atmosphere in the absence of solid carbon to suppress the chlorination of Si and vaporized the chlorides of other metals. Aluminum and Ti were selectively separated and recovered by condensation from the vapors. Overall, this method recovered 70 %–80 % Al and 80 % Ti. However, this method can become very expensive if the chlorination of unwanted metals is not suppressed.

Another method for the recovery of Al and Ti involves leaching CFA by acid/alkali leaching followed by precipitation, solvent extraction, or re-crystallization (Matjie et al. 2005; Halina et al. 2007; Li et al. 2011). US Patent No. 4243640 (1981) investigated a process for the recovery of Al from CFA where the ash was subjected to a magnetic separation and subsequently leaching was carried out on the non-magnetic fraction using nitric acid (strength about 30 %–56 %). The leachate was then evaporated and calcinated at 1000 °C for the recovery of Al. The calcinated product contained 90 % alumina along with other impurities like Fe oxides. US Patent No. 4567026 (1986) reported another acid leaching method for the recovery of Al and Ti from CFA. The process involved separation of magnetic fractions from the ash, followed by leaching the non-magnetic fraction with sulfuric acid, precipitating and removing Ti and Fe oxides from the leach solution by adding a strong base (NaOH), and precipitating and removing Al by contacting the reclaiming solution with $CO_2$. The precipitated aluminium hydroxide is calcinated to get aluminum oxide. However, direct sulfuric acid leaching yielded a low aluminum extraction (Matjie et al. 2005). Nayak and Panda (2010) also showed that direct sulphuric acid leaching at low concentration did not yield high aluminum recovery.

In contrast, CFA was treated pelletized with lime and calcinated, leaching with sulfuric acid that yielded better recovery. US Patent No. 4254088 (1981) discusses a process where the CFA was sintered with a mixture of NaCl and $Na_2CO_3$ at a temperature 700–900 °C. After sintering, 90 % of the aluminum CFA was in the acid soluble fraction. Recently, Matjie et al. (2005) discussed this calcinated technique at high temperature to decompose mullite phases, and then the calcinated CFA was leached in sulfuric acid to produce a solution containing Fe, Al and Ti species. The selective recovery of Al and Ti was carried out using various extraction methods, solvent extraction was more efficient than others and recovered 99.4 % of $Al_2O_3$ and 97 % of $TiO_2$. US Patent No. 4539187 (1985) also used an acid treatment where CFA was reacted with concentrated fluorosilicic acid and hydrogen fluoride at high temperature to form aqueous silicon fluoride vapor and fluoride and fluorosilicate of Al and Fe. Separated the Al and Fe bearing fluorides and fluorosilicate from aqueous silicon vapor, hydrolyzed and then recovered the substantial amount of pure Fe and Al following electroplating process. Recently, Wu et al. (2012) demonstrated pressure acid-leaching method, wherein the high extraction (82.4 %) of Al was obtained by diminishing CFA size (74 μm) and increasing acid concentrations (50 %). However in this method, the non-target metals readily leached into the acid, resulting in impurities. In addition, acid leaching had limited application due to its highly corrosive nature, besides this process required high cost, acid resistant equipment, and air-tight processing to protect the work environment.

Recovery of alumina by alkaline leaching is less problematic than techniques using acid leaching. US Patent No. 4130627 (1978) describes the alkaline leaching method where in the process takes place in an aqueous alkali solution of approximately pH 13, using NaOH, to form an aqueous CFA blend. The blend was agitated on a hotplate at temperatures of about 90 °C, and filtered to recover an alkaline solid. The filtrate was then contacted with hydrochloric or hydrofluoric acid to form soluble salts of various metals. Iron and aluminum were then selectively precipitated and recovered. This process recovered 65 % by weight of Fe and about 89 % of Al. However, this process has some disadvantages, such as being complex, time consuming, requiring larger quantities of processing materials, and high facility costs. This problem can be dealt with lime-soda sinter process, where a mixture of CFA, calcium carbonate, and sodium carbonate was sintered at high temperature to form a soluble sodium and calcium aluminate (Padil and Sohn 1985). Lu et al. (2003) reported this process where the CFA was mixed with limestone, sintered at 1260 °C, and then dissolved in sodium carbonate solution at 80 °C. This process obtained 99.9 % purity of alumina, and its recovery efficiency was 70 %. Recently, Bai et al. (2010) also used a lime-soda sintering method for the recovery of aluminum from an alkali desilicated CFA. The final Al extraction efficiency was 90 %. This process extracted Al from nearly 88 %–92 % at temperatures below 85 °C, while Al was not fully released from the sintered ash when the temperature was greater than 85 °C. US Patent No. 4397822A (1983) demonstrated an improvement method which added a small amount of sulfur and carbon to the lime-CFA sinter mixture before sintering to lower the optimum sinter temperature. This method recovered almost 90 % of alumina from CFA, and the results indicate that the addition of a combination of S and C, either elementally or as coal refuse, produces a higher recovery of alumina from CFA and at lower sintering temperatures, resulting in an energy savings. Recently, two US Patent Nos. 0119426A1 (2010) and 7871583B2 (2011) released a method for the recovery of alumina from high Si containing CFA, where CFA is leached with highly concentrated NaOH (>40 %). Silica is first removed from the CFA in the form of $Na_2SiO_3$, followed by separation of the residues having Al:Si ≥ 2. Subsequently the product was calcinated at 900–1100 °C to get pure alumina.

Besides acid/alkaline leaching, bioleaching also is used for the recovery of metals from CFA. This is discussed by in US Patent No. 5231018 (1993), where the CFA is suspended in an aqueous saline solution having pH ranging from 5 to 8, and then forming a reactive mixture by adding a pure culture a gram-positive bacteria (GIN-1), genus Rhodococcus. The mixture is then incubated for a period of time sufficient for the formation of microorganism cells/metal oxide adsorbate agglomerates, which separate from the reaction mixture. In this process, incubation reaction time was adjusted to selectively recover $TiO_2$ and $Al_2O_3$ from CFA. In another biochemical study, US Patent No. 5278069 (1994) used *Thiobacilus thiooxidans* strain in an acidic media for the recovery of metals from CFA. The metals were selectively recovered by precipitating the solution by adjusting the pH. Aluminum was mostly precipitated at pH 3.1–4.7, while Ti was precipitated at pH 1.5–3.1. These results indicate that bioleaching is one of the most cost-effective methods for the removal of metals from CFA; however, the percentage of metal recovery and their purity have not been reported.

## 3.2 Silica

Silica is used for various purposes, such as semi-conductors, soda-lime glass, high temperature thermal protection fabrics, glasses (including borosilicate glass and lead glass), and contact lenses and medical devices (Lukkassen and Meidell 2007). Coal fly ash is enriched with silica.

Several methods have been developed for the recovery of silica such as acid or alkali leaching, followed by separation and purification using different techniques. US Patent No. 4539187 (1985) describes a method where CFA is reacted with fluorosilicic acid and hydrogen fluoride at a high temperature to form fluorides of Fe, Al and Si. The fluoride mixture is distilled, separating the aqueous and gaseous silicon fluoride from the mixtures, hydrolyzing the silicon fluorides vapor, and recovering silicon dioxide in high purity form. Recently, US Patent No. 0287827A1 (2010) released a method where a mixture of coal and CFA is reacted with hydrogen fluoride in water to produce a liquid stream containing silicon fluoride and metal fluorides, and a solid stream containing unreacted coal and S compounds. Sulfur compounds are reacted with metallic nitrates dissolved in the water, which separated the aqueous solution of nitrate, S and metallic ions from the solid coal, and the solid coal is then washed with water. Subsequently, silicon fluoride and metal fluorides are contacted with metallic nitrates in an aqueous mixture to form solid silicon dioxide, and solid silicon dioxide is separated from the aqueous mixture. This process has recovered high purity silica from CFA. However, acid leaching is not always appropriate because it needs better acid resistant and air-tight processing equipment and it causes environmental pollution. This problem can be solved by using alkaline leaching methods. US Patent Nos. 0119426A1 (2010) and 7871583B2 (2011) released an alkaline leaching method for the recovery of silica from CFA, where Si is leached using NaOH in the form of $Na_2SiO_3$ solution. The solution is vaporized to obtain $Na_2SiO_3$, and pure silica (>98.5 %) is produced by carbonation.

# 4 RECOVERY OF TRACE ELEMENTS

There are a number of methods have been developed for the recovery of trace elements from CFA (Table 4), which are discussed below.

## 4.1 Vanadium

Vanadium (V) is mainly used in ferrous and non-ferrous alloys due to its tensile strength, hardness, and fatigue resistance (Moskalyk and Alfantazi 2003; Navarro et al., 2007). In addition, V is used in vanadium redox batteries. Although, V is extracted from various sources, V-enriched coal and carbonaceous residues such as CFA are one of the most important sources (Dai et al. 2015a).

Numerous methods have been used to extract V from carbonaceous residues. The most common method is wet treatment where the residue is subjected to multi-stage leaching, followed by precipitation via pH adjustment and oxidation/reduction (Europian Patent No. JP-A-H08-325651 1996; H10-251025 1998). US Patent No. 4814150 (1989) created a process where the combustion residue is reacted with sodium hydroxide in the presence of an oxidizer, dried at 100–250 °C, and leached with water to selectively extract V. Vanadium is precipitated from the alkaline extract by adding $H_2SO_4$ (up to pH 2) and an ammonium salt. The precipitated ammonium polyvanadate is calcinated to get vanadium pentoxide, which yields 80 %–90 % of V. US Patent No. 4966761 (1990) discusses a process in which CFA is mixed with water to prepare a slurry, carbon is skimmed off followed by alkaline pressure leaching and solvent extraction for the recovery of V (up to 93.4 %). US Patent No. 7498007B2 (2009) uses an acid solution to recover V. The extraction is carried out by precipitating vanadium pentoxide as a result of increasing pH by adding calcium hydroxide, quick lime, or calcium carbonate to the acid solution. After V is precipitated, the solid is separated by filtration or centrifugation. The percentage of V recovery from the acidic solution was ≥98 %. Recently, Xiang-yang et al. (2010) also described an acid leaching process using large volumes of sulfuric acid, extracting V up to 95.8 % from the leachate; however, no purification of V was attempted. Wet treatment techniques using acid/alkali leaching still have problems since they need a relatively large amount of expensive reagents, and they deal with complicated, troublesome repetition of steps, including conversion of solid CFA to a solution, isolating V from other metals and reconversion of V the solution into a solid. In addition, this technique may be unsuitable for CFA samples containing a high carbonaceous fraction and a low concentration of V because leaching V may not be efficient. To cope with this problem, dry treatment techniques have been developed (US Patent Nos. 4420464 1983, 4816236 1989). In these methods, the carbon is removed by combusting the V-containing residue at a controlled temperature, and then the residue is leached with a strong alkali/acid to recover the vanadium. Vitolo et al. (2000) used this method for the recovery of V from a fly ash at temperatures of 650–1150 °C, followed by an acid leaching and oxidative precipitation of vanadium pentoxide. Overall, this process yielded 83 % V. US Patent No. 6652819B2 (2003) also employed this method for the recovery of V from carbonaceous residues. The method involves a preliminary combustion of the residue at temperatures of 500–690 °C, followed by leaching with $H_2SO_4$ at a pH range of 1.5–4, and selective precipitation V by adjusting the pH in the range of 4.5–7.5, and separating the V ion as a tetravalent vanadium compound.

## 4.2 Gallium and germanium

Gallium (Ga) is used for the manufacture of high technology optical devices such as semiconductors, DVD's, laser diodes, and other electronic devices (Moskaly 2004). Germanium (Ge) is used in the manufacture of photovoltaic cells, fiber optics and as a catalyst for polyethylene terephthalate production (Butterman and Jogerson 2005). Germanium and gallium in the Earth's crust are mainly associated with Zn and its mineral phases. Thus, they are typically recovered from by-products of Zn mineral processing. Germanium and Ga are also enriched in coals (Gluskoter et al. 1977; Harvey et al. 1983; Dai et al. 2012a, b; Seredin 2012; Dai et al. 2014b; Qin et al. 2015a, b; Dai et al. 2015b). Thus, coal fly ash is a promising alternative resource of those elements (Fang and Gesser 1996; Zhang et al. 1999; Arroyo et al. 2014; Dai et al. 2014a, b). After combustion in a coal burning power plant, both elements condense from the vapor phase onto the surface of the CFA, thus are enriched in CFA relative to the feed coal (Font et al. 2005; Querol et al. 1995).

Several techniques have been developed for the recovery of Ge and Ga from CFA. Alkali halite melt is one of the methods, and is discussed in US Patent No. 4475993 (1984). In this method, the CFA is reacted with aluminium trichloride to form a soluble compound of trace metals in the melt. Gallium was recovered from the melt by electrolysis. However, this process is uneconomical and recovered a moderate percentage (69 %) of Ga. Another method of Ga and Ge extraction is sublimation, which is discussed in several US Patents (Nos. 4757772 1988, 4757770 1988, and 4643110 1987). The steps include heating the CFA in an oxidizing atmosphere to vaporize the Ge and Ga, then treating the oxides in a reducing atmosphere to reduce oxides to suboxides, which subsequently precipitate on the surface of sand particles, and recovery of the products from the sand particles using a cold liquid solution. This method has problems involving other volatile compounds such as As and S. To cope with this problem, US Patent No. 4678647 (1986) introduced a method where CFA is pelletized and treated in the presence of an oxidizing gas at a temperature of about 900 °C. This treatment removes As and S from the CFA, so that they will not remain in the concentrated Ga and Ge fractions. Then, the oxides are treated in the presence of a reducing gas at the same temperature range to reduce the oxides to suboxides. Recover of Ga suboxides is done by condensation at a lower temperature (nearly 700–800 °C) and dissolving the condensed material using an acidic solution. The recovery rate of Ga for this method is up to 90 %. This method can be more economical and more efficient than the alkali halite method.

Other techniques for the recovery of Ga and Ge include using water/acid/alkali leaching of CFA, followed by separation/purification by using precipitation, distillation, ion flotation, adsorption into activated carbon, and solvent extraction (Meawad et al. 2010). Arroyo (2014) optimized Ge and Ga recovery from CFA using different extractants in a wide range of extraction conditions (acid, alkaline, complexant, reducing and oxidizing). High extraction yields of Ge (up to 90 %) and Ga (up to 82 %) were made using weak oxalic acid and sulfuric acid solution, respectively within 1–2 h extraction period. Gutiérrez et al. (1997) discuss the recovery of Ga from CFA by 6 M hydrochloric acid in contact with Amberlite LA-2. Iron is precipitated with sodium hydroxide, while 83 % of Ga remains in solution, which is extracted selectively with LIX 54. Fang and Gesser (1996) studied the recovery of Ga from CFA involving a of two stage acid leaching with hydrochloric acid, followed by removal of silica and calcium, reduction of iron, extraction of Ga by foam floatation (polyurethane foam) and finally purification. This method showed that better extraction of Ga occurs at higher acid concentrations, low temperatures, short extraction times and smaller particle size. However, acid leaching is costly, requires better acid resistant equipment, and better environmental protection. Arroyo et al. (2009) discussed a water leaching process for extracting Ge from CFA, followed by solvent extraction. In this process, Ge complexation achieved with catechol, then extraction of the Ge-catechol complex with an organic reagent such as trioctylamine diluted in an organic solvent, and finally, stripping Ge from the organic extract by NaOH. This process extracted 95 % of the Ge from the leachate. In another study, Font et al. (2005) invented a low cost and environment friendly process for the recovery of Ge, where the CFA is leached in water at different ash/water ratios and temperatures. Results showed that temperature is the most important parameter in the extraction process, maximum content (up to 86 %) of the Ge extracted at 90 °C. For the recovery of Ge from solution, Hernandez-Exposito et al. (2006) developed an ion flotation method using different complexing agents (pyrogallol, catechol, hydroquinone and resorcin) at pH range of 4–7. It was found that using dodecyl amine as a surfactant, and pyrogallol or catechol as a complexing agent at pH values of 4–7, 100 % of the Ge could be recovered in 30 min. Torralvo and Fernández-pereira (2011) also reported that catechol as complexing agent for the recovery of Ge from leachates.

All the above methods were tested on bulk CFA; however, Ge and Ga are mainly concentrated in the finer fractions. This was examined in US Patent No. 4686031 (1987), where the process is mainly subjected to size classified CFA, and then extraction by various treatments. It was found that fine fraction, isolating 30 % of the finest particles while avoiding substantial rupturing of cenospheres and plerosphere, provided the best recovery.

### 4.3 Molybdenum

Molybdenum (Mo) is used to produce high tensile strength steel. It is also used as a biocatalyst for plants and animals and it is widely used in the manufacture of super alloys, lubricants, and electronic products. Coal fly ash can provide an alternative source for Mo recovery since some coals are enriched with Mo (Dai et al. 2015a, b). Mok et al. (1984) extracted Mo from CFA using the steps, digestion of CFA using concentrated HCl and $HNO_3$, diluting the solution using HCl, and selectively extracting Mo from the solution using dithiocarbamate. Ogata et al. (2011) disclosed a method for the recovery of Mo from CFA where the CFA is leached at pH levels 2, 7 and 11, and then adjusted with hydrochloric acid or sodium hydroxide solution. Subsequently, gibbsite (GB), an amorphous aluminium oxide, is used for the adsorption of Mo from elutes from the solutions. A greater concentration of Mo was eluted under acidic conditions onto GB400 (gibbsite calcinated at 400 °C) at pH 2 than from neutral or basic eluates. Recovery of Mo from eluates was 90 % using GB400. The main adsorption mechanism was ion exchange with a number of the hydroxyl groups in GB.

### 4.4 Lithium

There are a number of applications used for lithium carbonate, including as an additive in aluminum smelting, production of glasses, enamels, and ceramics, and production of electronic grade crystals. Its pure form is used in lithium ion batteries. Lithium has been found enriched in coal deposits (Qin et al. 2015b). US Patent No. 3857920 (1974) describes a method for the recovery of lithium carbonate from slurries resulting from the dispersion of CFA in water. In this method, CFA slurry is treated with carbon dioxide to convert the lithium carbonate to the more soluble lithium bicarbonate. The resulting bicarbonate solution is then filtered to remove insoluble material. The filtrate containing lithium bicarbonate is decarbonated either by vacuum treatment, heat, or preferably by aeration of the solution to precipitate lithium carbonate in the aqueous medium. The precipitated lithium carbonate is then filtered from the aqueous solution. This method is inexpensive and it recovered 80 %–90 % of the lithium carbonate.

## 4.5 Selenium

Selenium (Se) is mainly used in glassmaking and for pigments. It is also used for making semiconductors and in photocells. Selenium is commercially extracted as a byproduct in the refining of metal sulfide ores, most often during copper production. It can also be extracted from CFA since some coals are enriched in Se (Dai et al. 2014a, b; 2015b, c). Wang et al. (2007) studied the leaching of selenium from bituminous and sub-bituminous coal ash with respect to pH. Very low concentrations of selenium were released at a pH range from 3 to 4, while the amount increased with increasing pH above 4, and the maximum amount of Se (55 %–69 % of the total) was leached at pH 12. Iwashita et al. (2005) also reported that leaching of Se increased when pH was raised. However, recovery of this element is not attempted in these literatures.

## 4.6 Radioactive elements

Radioactive elements have many industrial and scientific applications, despite the obvious dangers if they are not handled properly. Radioactive elements are used as tracers in the medical field, the generation of electricity in nuclear power plants and in the production of nuclear weapons. Coal contains significant amounts of radioactive elements such as uranium (U) (Dai et al. 2015a, c). Thus, CFA can be alternative source of these elements. Sparton Resources Inc. (2007) reported the extraction of U from CFA. This process included acid leaching ($H_2SO_4$) and formation of yellowcake by passing the filtered leachate solution through a standard ion exchange resin, stripping the resin and precipitating a U compound. This process is essentially similar to the U extraction method used by the primary U ore processing plants. However, this work is preliminary and requires further work to improve recovery.

Maslov et al. (2010) discussed a method for recovering U from brown coal ash of Mangolia where the ash is leached with an acid mixture (8 M $HNO_3$ + 10 % HF). The acid mixture leached U and Ra as high as 99 % and 97 %, respectively. The extraction of U from the solution and its purification for the removal of accompanying elements were carried out using an anion exchanger (Dowex 2X8) with a 2 M HCl solution as the eluate. After leaching, the CFA contained neither natural radioactive elements nor their decay products, thereby allowing for the use of the ash as a building material.

## 4.7 Gold, silver and platinum group of elements

Precious metals such as gold (Au) and silver (Ag) have many application in industry, jewelry and medicine (Guerra and Calligaro 2003; Drake and Hazelwood 2005; Eisler 2003). Some Au compounds are used as anti-inflammatory drugs in the treatment of rheumatoid arthritis and Ag compounds are used in the treatment of infections. Gold is also found in electronic apparatus and Ag is used in photography, industry and for hard solder.

US Patent Nos. 6827837B2 (2004) and 0056548A1 (2005) released a method for the recovering of gold and one and more platinum group elements, such as platinum, rhodium, ruthenium, osmium and iridium, from CFA. In this method an ash formed by the combustion of a pre-selected coal is mixed with the ash contains gold and one more platinum group of elements, an inquart containing gold and one or more platinum group elements, and fluxing agent to form a 'charge'. The 'charge' is then heated in a furnace to a preselected temperature and forming a molten charge including a molten inquart; absorbing a portion of the gold and more platinum group elements of the ash into the molten inquart. Casting a dare bar (which contains gold and one or more platinum group elements) by discharging the molten inquart. Digesting the dore bar in an elctrowinning process where the dore bar is introduced as an anode in an acidic bath and an electrical potential is maintained between the cathode and anode, then precipitating and dissolving trace elements in the electrolytic solution producing an anode slime. The digestion of anode slimes is done using 30 % solution of $HNO_3$. The digestion is filtered and rinsed then the liquid rinsed portion containing Ag in solution is washed with HCl and filtered. The residue is mixed with aqua regia, forming a solution containing gold, platinum and palladium. Gold may then be removed by solvent extraction. Rhodium and other metals may be removed by precipitation by the treatment of the solution with potassium hydroxide and sodium borohydrite. Palladium in solution may be removed by precipitation by treatment with ammonium hydroxide, $NH_4OH$, and HCl as the complex $PdCl_2(NH_3)_2$. Alternatively, Pd in solution is removed by precipitation with DiMethyl Glyxime, followed by filtration and reduction and ashing of the residue. Platinum in solution is removed by any of the accepted methods including precipitation as impure $(NH_4)_2PtCl_6$ after treatment with $NH_4Cl$, or leaching $H_2PdCl_4$ from solution. An organic solvent such as DiMethyl Glyoxime is used to extract gold, after treatment with $NH_4Cl$, the solution should also be treated with ethanol and hydrogen peroxide. The solution is filtered and the residue, including $(NH_4)_2PtCl_6$, subjected to reduction and burning leaving an impure platinum sponge. The platinum sponge can be purified by redissolving in aqua-regia. This method is very complex, and a recovery and purity level is not mentioned.

## 4.8 Rare earth elements (REEs)

Rare earth elements include light REEs (Sc, La, Ce, Pr, Nd, and Pm), medium REEs (Sm, Eu and Gd) and heavy REEs (Tb, Dy, Ho, Er, Tm, Yb, Lu and Y), which are mainly used is various industrial purposes such as fuel cells, green energy devices, hi-capacity batteries, magnets for wind generation (Chen 2011; Hoenderdal et al. 2013; Franus et al. 2015).

REEs have been recovered from bauxite residue (Borra et al. 2015) and electronic wastes (Anonymous 2015). It has also been recognized that some coals (Seredin 1996; Dai et al. 2014a, b) and CFA (Ketris and Yudovich 2009; Blissett et al. 2014; Dai et al. 2012a, b; Dai et al. 2014a, b; Franus et al. 2015) are enriched with REE; thus, recently sparking interest in evaluating of this product as a potential alternative source of REEs (Seredin et al. 2013).

There are several methods developed for recovery of rare elements from CFA, the chlorination method is favored due to selectivity and high reactivity. The necessary apparatus required for this process is very expensive. This technique can be used to recover only certain rare elements where other complex recovery processes are not suitable. From this standpoint, US Patent No. 4649031 (1987) reported a non-destructive digestion process for the recovery of rare elements, including Lanthanides and Actinides and group V and VI elements that are adsorbed onto the surfaces of coal combustion residues. This techniques includes: (1) reducing the grain size of the combustion residues from 0.1 to 1 mm, (2) digestion of the residues with water, (3) digestion of the residues with a dilute aqueous alkali solution (1.5 %–8 %), (4) digestion of the residue with a dilute aqueous acid solution (0.5 %–5 %) at a temperature below the boiling point of water, (5) separating the solid and liquid phases after each digestion, (6) and isolating the s metals from the liquid phases. This study shows that the mild digestion agents used dissolve an unexpectedly high proportion of rare metals present in the byproducts and that increasing the heat and residence time lowers the availability of the metals. This process is not appropriate for the recovery of individual elements from the mixed solution.

Recently, US Patent No. 0287653A1 (2013) demonstrated the recovery of REEs from CFA. In this method, first the CFA was treated with a mineral acid ($HNO_3$) at 90 °C to form a more concentrated mineral acid solution. This solution is mixed with an organic solution (tributyl phosphate and kerosene) that includes the rare earth salts. The organic solution is mixed with water to form an aqueous solution. REEs substantially recovered from the aqueous solution by using ion exchange process. In a continuation-in-part of the above patent, US Patent No. 0139871 (2015) was carried out using the same method, but here the CFA was sorted into a substantially magnetic and non-magentic fraction. This indicates magnetic treatment can maximize efficiency of extracting REEs from CFA. However, extraction efficiency of these methods is not reported. Furthermore, ion exchange method can remove other metals from the solution. This method needs to be optimized to improve the yields for REEs.

## 4.9 Other trace elements (Pb, Cr, Cd, and Cu)

Coal fly ash also contains considerable amounts of trace and rare elements, and some them can be hazardous (Izquierdo and Querol 2012). There are many studies which focused on the extraction of trace metals from CFA, some employing hydrometallurgical techniques using various lixiviants: inorganic acids such as sulfuric acid, hydrochloric acid, nitric acid, organic acid such as acetic acid, citric acid, oxalic acid, alkaline reagents such as sodium and ammonium hydroxide, chelating reagents such as ethylendiaminetetraacetate (EDTA) and diethylenetriaminepentaacetate (DTPA) (Hong et al. 2000; Zhang and Itoh 2006).

Narukawa et al. (2007) studied Cr extraction from an Australian CFA using 0.01 M NaOH. However, this process did not leach Cr(III) from the CFA. Huang et al. (2011) discussed the alkaline leaching by ammonium or sodium hydroxide, where only Pb and Zn are recovered while other metals remained in the solid residues. However, this process can extract other heavy metals such as Cu and Cd if the leaching process is conducted together with subsequent leaching using acidic lixiviants. Okada et al. (2007) reported that in acidic leaching using acetic acid recovered 97 % of Zn and 98 % of Pb from CFA, while the same CFA in alkaline leaching using 3 M NaOH recovered 81 % Pb and 35.5 % Zn.

Consequently, acid leaching is more effective under hydrothermal conditions. This was studied by Zhang and Itoh (2006) for CFA using five types of acids and hydrochloric acid was found to be the most effective for metal extraction from the CFA. Compared to room conditions (standard temperatures and pressures), it was determined that hydrothermal treatment accelerated the dissolution of the CFA and resulted in better leaching of metals such as Cr, Cd, and Pb. Even more effective leaching can be done using chelating agents. Hong et al. (2000) studied the extraction of heavy metals from CFA using acid and chelating agents such as nitritriacetic acid (NTA) and ethylendiaminetetraacetate (EDTA), or diethylenetriaminepentacetate (DTPA). The results showed that chelating agents provide good leaching efficiency and after the treatment the CFA's were successfully detoxified to meet landfilling guidelines. Using 3.0 % EDTA or DTPA, 20 %–50 % Cr, 60 %–95 % Cu, 60 %–100 % Pb, and 50 % –100 % Zn were extracted from CFA. However, these agents are not biodegradable and it is difficult to recover the metals from the leachate due to their strong chelating properties.

Recovery of metals from CFA using bioextractants has been studied by many authors (Seidel et al. 2001; Ishigaki et al. 2005; Sierra-Alvarez 2009). This process is important at an industrial scale. In this process, chemolithoautotrophic bacteria such as Thiobacillus species and Heterotrophic fungi such as Aspergillus and Penicillium species are used to leachthe metals. The leached metals can then be recovered and the residues may be safely utilized. US Patent No. 5278069 (1994) discusses the bioextraction process using strains of Rhodococcus and *Thiobacillus thiooxidans* for the removal of Co along with other metals (Al, Ti and Fe) from CFA. After extraction, the metals were recovered in a series of precipitation steps by controlling pH. This technique is relatively inexpensive and requires low energy inputs when compared with conventional technologies (Krebs et al. 1997). However it is a slow process requires sufficient amounts of cell culture and is not selective toward specific metals (Groudev et al. 1982).

# 5 RECOVERY OF OTHER MATERIALS

## 5.1 Magnetic materials

Magnetite or other magnetic material is used by the steel industry for cleaning coal prior to combustion. One potential source of magnetite is CFA which can contains economically recoverable amounts of magnetic material. After removal of magnetic material from CFA, the remaining product can be useful for refractory and ceramic manufacturing since it contains appreciable amounts of $Al_2O_3$ and $SiO_2$ (Rao et al. 1999). The characteristics of magnetic material in CFA vary depending on the coal source and operating conditions of the coal boiler (Zyryanov et al. 2011). The basic principle behind magnetic separation deals with the fact that materials having different magnetic moments experience varying forces in the presence of a magnetic field gradient. Thus, applying an external magnetic field can extract those materials with similar magnetic properties (Yavuz et al. 2009).

There are many wet or dry magnetic separation methods for the recovery of magnetic materials from CFA. The dry process uses a rotating magnetic drum and initial separation produces middling and tailings which are discarded and a crude iron oxide fraction which must be further refined by passing it through another magnetic unit. The wet process involves pumping CFA slurry into a magnetic drum where the magnetic iron oxides and associated materials are collected and then the material is passed through a similar unit for further refining. In most of the plants using wet magnetic separation, the advantage as that the slurry can be directly fed into the feed tank. Prakash et al. (2001) demonstrated the use of a selective magnetic coating for the removal of iron from Talcher (India) CFA. They used sodium hexametaphophate (NaHMP) as a dispersing solution, followed by sodium oleate for the separation of iron from. The recovery rate for Fe was around 92 %. The wet/dry magnetic separations processes are documented by numerous US Patents (Nos. 4191336 1980; 4319988 1982; 5047145 1991; 5227047 1993). However, these methods are not cost-effective if the ash contains less than 2 % magnetite. In order to improve the process or more recovery, CFA containing sufficient amount of magnetic material (>2 %) need to be selected. Groppo and Honaker (2009) studied the economic recovery of magnetite material from CFA.

US Patent No. 3769054 (1993) released another method for the recovery of magnetic materials from CFA. In this method CFA is dry processed in an air classifier and then separated the magnetic separation. This process is expensive, and the recovered portion of magnetite often contains undesirable impurities. To resolve these problems, US Patent No. 4191336 (1980) provides an improved process for the recovery of high purity magnetite from CFA. This process involves making a water slurry of the CFA and then passing the concentrated slurry through the magnetic separators; magnetite is contained in the separated material. The separated magnetite is then pulverized to a size where 95 % of the magnetite will pass through a 325 mesh sieve. The crushed magnetite is passed through a third separator followed by purification and drying.

## 5.2 Cenospheres

Cenospheres are composed mainly of silica, alumina and iron oxides, and have unique properties such as sphericity, inert, lower density than water, and they are waterproof making them useful in a variety of applications. Generally, cenospheres make up 1 %–3 % by weight of the total CFA produced. They have valuable industrial applications, such as oil well drilling, ceramics, plastic, paint, coatings, polymer fillers, building products, buoyaancy, and as low dielectric constant substrates (Huo et al. 2009; Wang et al. 2011). Cenospheres are used for water purification such as the removal of excess fluoride from drinking water. In this process, magnesia loaded cenospheres are produced using a relatively simple wet impregnation method with magnesium chloride (Xu et al. 2011). Cenospheres are used as a carrier because of their low weight, sphericity, non-toxicity, and high strength. They are ideal for incorporation into materials such as silicone rubber in order to increase the conductivity. In addition, they enhance the rubber's suitability as an electromagnetic wave absorbing material, which can be used in electronic and radar applications (Pang et al. 2011).

There are several methods for the separation of cenospheres from CFA. Traditionally, cenospheres were removed from CFA via float-sink in settling lagoons. This simple procedure allows potentially toxic elements to be leached and requires a large space and significant time (Hirajima et al. 2010). Alternatively, a triboelectric separation system was used (Gurupira et al. 2001) but this technique requires that the spheres have a specific gravity ≤1. Some solid silica particles with specific gravity of ≥1 may contain gas bubbles which over their specific gravity so that they mimic cenospheres during this process, rendering it less effective and attractive. Other studies that have looked at the extraction of cenospheres on a wet basis. In US Patent No. 4121945 (1978), CFA is mixed with water to form a slurry then a frothing agent is added and cenospheres lighter than water are skimmed. US Patent No. 4652433 (1987) looked at another method where CFA is separated into fine and coarse fractions and the coarse fraction is then mixed with water, subsequently the cenospheres are removed by skimming. These methods have some disadvantages. Wet processing caused environmental problems and imposed space restrictions. In addition, these methods only collect cenospheres with specific gravity ≤1.0; those that float on water. Also, these processes are not suitable for CFA that contains 10 % or more lime because high Ca bearing CFA (such as ASTM C618 Class C fly ash) has cementations properties when mixed with water and when pozolanic reactions occur the material may harden. When using these wet methods many cenospheres can be trapped in agglomerated and/or hardened masses before flotation occurs. This problem was solved by a few recent US Patents (No. 0190327A1 2008, No. 8074804B2 2011, and No. 0050519A1 2012) where cenospheres were separated from CFA in a dry state by using air as the fluid media for the separation. In one of the processes, CFA is separated into narrow particle size fractions, followed by density separation using a fluidized bed of air, preferable in the absence of water by dry screening, and then an optional further particle size separation. The advantages of this technique are that it can yield nearly 100 % cenospheres and it avoids agglomeration of ash having high Ca content.

### 5.3 Unburned carbon

Unburned carbon in CFA can be used during activated carbon preparation or carbon black, and can be used as a supplementary fuel. It can also be useful in the manufacture of graphite (Cabielles et al. 2008; Cameán and Garcia 2011). Removal of unburned carbon from coal ash makes it a more valuable cement additive. There are numerous methods for the recovery of unburned carbon from CFA (Hwang et al. 2002). Electrostatic separation is one of them, has been used by many authors (Gray et al. 2002; Soong et al. 2002). This method operates in a controlled electrode potential condition following the principle of bipolar charging of dry particles either by particle–particle contact or by particle–wall collision under turbulent conditions. This process separates the positively charged carbon particles from the negatively charged ash particles. This technology has been successfully commercialized.

One of the most promising methods for recovering unburned carbon from CFA is froth flotation. This method relies on the ability of air bubbles to selectively adhere to the surface of a particular mineral surface. Since carbon is hydrophobic in nature, other particles that are hydrophilic attached to the air bubbles and become positively buoyant and float on the slurry (Ucurum 2009). This process has been commercialized in the US for CFA beneficiation with reported carbon grade recoveries of up to 70 % (Baltrus et al. 2001).

A similar technology for froth flotation uses oil agglomeration, which involves wetting hydrophobic particles by adding oil to the aqueous slurry. Subsequently, the vessel can be agitated which allows the oil coated particles to collide with each other and form agglomerates. The oil/carbon agglomerates being less dense than the ash suspension rise to the top of the vessel (Mehrotra et al. 1983). US Patent No. 546363 (1995) removes carbon adding floating reagent-mixture of fuel oil and petroleum sulfonate (10:1) and has recovered 44 % after 1 min of flotation while fuel oil flotation after 20 min recovered only 38 %. Gray et al. (2001) used this method with cyclohexane as the solvent, and achieved high carbon recoveries (55 %–57 %). There are other studies reporting similar carbon purities in their agglomerates using vegetable oil as their solvent (Alonso et al. 1999). US Patent No. 6068131 (2000) describes this type of removal using a conditioning agent such as biodegradable soil.

# 6 CONCLUSIONS

Coal fly ash has been explored as an alternative source of various valuable elements and minerals. Recovery of these components from CFA not only provides alternative resources, but also reduces disposal cost and protects the environment. This paper has reviewed a number of hydrometallurgical methods developed for the recovery of those products from CFA and highlighted suitability and effective technique for their extraction based on US Patents and journal literature. By applying those techniques, it was found that the recovery of major elements such as Fe, Al and Si and trace elements such as Ge and Ga from CFA are satisfactory. However, for other trace and rare earth elements further research is needed to develop adequate results in terms of selective recovery and purification. Mineral matter such as cenospheres, unburned carbon and magnetic materials also significantly extracted from CFA.

Most of these methods were studied at lab-scale, their commercialization utilization is largely limited, and therefore further development in large scale is required to scale up to commercial production levels. In addition, economic viability and environmental impact evaluation of each method should be focused in the future work. This would be helpful to choosing correct method and to ensure that any operations don't result in a greater environmental footprint than mining of primary resources. Quality of raw ash obtained in parallel with the metal recovery processes needs to be assessed.

## REFERENCES

1. Ahmaruzzaman M (2010) A review on the utilization of fly ash. Prog Energy Combust Sci 36:327–363

2. Alam J, Akhtar MN (2011) Fly ash utilization in different sectors in Indian scenario. Int J Emerg Trends Eng Dev 1:1–13

3. Alonso MI, Martínez-Tarazona RM, García AB (1999) Coal recovery from coal fines clearing wastes by agglomeration with vegetable oils: effect of oil type and concentrations. Fuel 78:753–759

4. Aldrich RG (1982) Separation of high grade magnetite from fly ash. US Patent No. 4319988

5. Anonymous (2015) Recovery of rare earths from electronic wastes; an opportunity for High-Tech SMEs. Directorate General for Internal Policies, Policy Department A: Economic and Scientific Policy, European Parliament. http://www.europarl.europa.eu/RegData/etudes/STUD/2015/518777/IPOL_STU(2015)518777_EN.pdf)

6. Arroyo F, Fernandez-Pereira C, Olivares J, Coca P (2009) Hydrometallurgical recovery of germanium from coal gasification fly ash. Ind Eng Chem Res 48:3573–3579

7. Arroyo F, Font O, Chimenos JM, Pereira CF, Querol X, Coca P (2014) IGCC fly ash valorisation. Optimisation of Ge and Ga recovery for an industrial application. Fuel Process Technol 124:222–227

8. Ashworth RA et al. (1987) Method for the recovery of minerals and production of by-products from coal ash. US Patent No. 4652433

9. Bai G, Teng W, Wang X, Qin J, Xu J, Li P (2010) Alkali desilicated coal fly ash as substitute of bauxite in lime-soda sintering process for aluminium production. Trans Nonferr Met Soc China 20:169–175

10. Baltrus JP, Wells AW, Fauth DJ, Diehl JR, White CM (2001) Characterization of carbon concentrates from coal-combustion fly ash. Energy Fuels 15:455–462

11. Basu M, Pande M, Bhadoria PBS, Mahapatra SC (2009) Potential fly-ash utilization in agriculture in agriculture: a global review. Prog Nat Sci 19:1173–1186

12. Barclay KM (1983) Recovery of vanadium from carbonaceous materials. US Patent No. 4420464

13. Benson S, Laumb M (2007) Removal and recovery of deposits from coal gasification systems. US Patent No. 20070274886A1

14. Blander M et al. (1984) Extraction of trace metals from fly ash. US Patent No. 4475993

15. Blissett RS, Rowson NA (2012) A review of the multi-component utilization of coal fly ash. Fuel. doi:10.1016/j.juel.2012.03.024

16. Blissett RS, Smalley N, Rowson NA (2014) An investigation into six coal fly ashes from the United Kingdom and Poland to evaluate rare earth element content. Fuel 119:236–239

17. Borra CR, Pontikes Y, Binemans K, Gerven TV (2015) Leaching of rare earths from bauxite residue (red mud). Miner Eng 76:20–27

18. Burnet O, Murtha MJ, Wijatno H (1977) Recovery of alumina from fly ash by high temperature chlorination. In: Rose JG (ed) Proceeding 3rd Kentucky coal refuse disposal and utilization seminar, May 11–12. University of Kentucky, Lexington, pp 83–88

19. Butterman WC, Jogerson JD (2005) Mineral commodity profiles: germanium. U.S. Geological Survey, U.S. Department of the Interior, Washington, DC

20. Brown JW (1980) Process for recovering magnetite from fly ash. US Patent No. 4191336

21. Cabielles M, Montes-Morán MA, Garcia AB (2008) Structural study of graphite materials prepared by HTT of unburned carbon concentrates from coal combustion fly ashes. Energy Fuels 22:1239–1243

22. Cameán I, Garcia AB (2011) Graphite materials prepared by HTT of unburned carbon from coal combustion fly ashes: performance as anodes in lithium-ion batteries. J Power Sources 196:4816–4820

23. Cao D, Selic E, Herbell JD (2008) Utilization of fly ash from coal-fired power plants in China. J Zhejiang Univ 9:681–687

24. Chandrashekhar S (2010) Process for obtaining treated coal and silica from coal containing fly ash. US Patent No. 2010/0287827A1

25. Chen Z (2011) Global rare earth resources and scenarios of future rare earth industry. J Rare Earths 29:1–6

26. Clark LB, Sloss LL (1992) Trace elements—emissions from coal combustion and gasification. IEA Coal Research Rep. No. IEACR/49

27. Corigliano F et al. (1989) Process for the high-yield recovery of vanadium from petroleum combustion residues. US Patent No. 4814150

28. Dai S, Jiang Y, Ward C, Gu L, Seredin VV, Liu H, Zhou D, Wang X, Sun Y, Zou J, Ren D (2012a) Mineralogical and geochemical compositions of the coal in the Guanbanwusu Mine, Inner Mongolia, china: further evidence for the existence of an Al (Ga and REE) ore deposit in the Juangar Coalfield. Int J Coal Geol 98:10–40

29. Dai S, Wang X, Seredin VV, Hower JC, Ward CR, O'Keefe JMK, HuangW Li T, Li X, Liu H, Xue W, Zhao L (2012b) Petrology, mineralogy, and geochemistry of the Ge-rich coal from the Wulantuga Ge ore deposit, Inner Mongolia, China: new data and genetic implications. Int J Coal Geol 90–91:72–99

30. Dai S, Luo Y, Seredin VV, Ward CR, Hower JC, Zhao L, Liu S, Zhao C, Tian H, Zou J (2014a) Revisiting the late Permian coal from the Huayingshan, Sichuan, southwestern China: enrichment and occurrence modes of minerals and trace elements. Int J Coal Geol 122:110–128

31. Dai S, Seredin VV, Ward C, Jiang J, Hower J, Song X, Jiang Y, Wang X, Gornostaeva T, Li X, Liu H, Zhao L, Zhao C (2014b) Composition and modes of occurrence of minerals and elements in coal combustion products derived from high-Ge coals. Int J Coal Geol 121:79–97

32. Dai S, Seredin VV, Ward CR, Hower JC, Xing Y, Zhang W, Song W, Wang P (2015a) Enrichment of U–Se–Mo–Re–V in coals preserved within marine carbonate successions: geochemical and mineralogical data from the late Permian Guiding Coalfield, Guizhou, China. Miner Depos 50:159–186

33. Dai S, Liu J, Ward CR, Hower JC, Xie P, Jiang Y, Hood MM, O'keefe JMK, Song H (2015b) Petrological, geochemical, and mineralogical compositions of the low-Ge coals from the Shengil Coalfield, China: a comparative study with Ge-rish coals and a formation model for coal-hosted Ge ore deposit. Ore Geol Rev 71:318–349

34. Dai S, Yang J, Ward CR, Hower JC, Liu H, Garrison TM, French D, O'Keefe JMK (2015c) Geochemical and mineralogical evidence for a coal-hosted uranium deposit in the Yili Basin, Xinjiang, northwestern China. Ore Geol Rev 70:1–30

35. Dhadse S, Kumari P, Bhagia LJ (2008) Fly ash characterization and utilization and government initiatives in India: a review. J Sci Ind Res 67:11–18

36. Drake PL, Hazelwood KJ (2005) Exposure-related health effects of silver and silver compounds: a review. Ann Occup Hyg 49:575–585

37. Eisler R (2003) Chrysotherapy: a synoptic review. Inflamm Res 52:487–501

38. Fang Z, Gesser HD (1996) Recovery of gallium from coal fly ash. Hydrometallurgy 41:187–200

39. Fass R et al. (1994). Bioleaching method for the extraction of metals from coal fly ash using thiobacillus. US Patent No. 5278069

40. Font O, Querol X, Lopez-Soler A, Chimenos JM, Fernandez AI, Burgos S, Garcia PF (2005) Germanium extraction from gasification fly ash. Fuel 84:1384–1392

41. Fontana D, Kulkarni P, Pietrelli L (2005) Extraction of titanium (IV) from acidic media by 2-ethylhexyphosphonic acid mono-2-ethylhexyl ester. Hydrometallurgy 77:219–225

42. Franus W, Wiatros-Motyka MM, Wdowin M (2015) Coal fly ash as a resource for rare Earth elements. Environ Sci Pollut Int 22:9464–9474

43. Gardner HE (1989) Recovery of vanadium and nickel from petroleum residues. US Patent No. 4816236

44. Gieré R, Carleton LE, Lumpkin GR (2003) Micro- and nanochemistry of fly ash from a coal-fired power plant. Am Miner 88:853–1865

45. Gluskoter HJ, Ruth RR, Miller WG, Cahill RA, Dreher GB, Kuhn JK (1977) Trace elements in coal: occurrence and distribution. Illinois State Geological Survey, Circular 499

46. Grantham L et al (1974) Recovery of lithium carbonate. US Patent No. 3857920

47. Gray ML, Champagne KJ, Soong Y, Finseth DH (2001) Parametric study of the column oil agglomeration of fly ash. Fuel 80:867–871

48. Gray ML, Champagne KJ, Soong Y, Killmeyer RP, Maroto-Valer MM, Anderson JM, Ciocco MV, Zandhuis PH (2002) Physical cleaning of high carbon fly ash. Fuel Process Technol 76:11–21

49. Gropp JG et al. (1995) Method of removing carbon from fly ash. US Patent No. 546363

50. Groppo J, Honaker R (2009) Economic recovery of fly ash derived magnetics and evaluation for coal cleaning. In: World of coal ash conference, Lexington

51. Groudev SM, Gencev FN, Groudeva VI (1982) Use of microorganisms for recovery of aluminum from alumino silicate. Achiev Prospect Travaux 12:203–212

52. Guerra MF, Calligaro T (2003) Gold cultural heritage objects: a review of studies of provenance and manufacturing technologies. Meas Sci Technol 14:1527–1537

53. Gurupira T, Jones CL, Howard A, Lockert C, Wandell T, Stencel JM (2001) New products from coal combustion ash: selective extraction of particles with density less than 2. In: International ash utilization symposium, Lexington, Kentucky

54. Gutiérrez B, Pazos C, Coca J (1997) Recovery of gallium from coal fly ash by a dual reactive extraction process. Waste Manag Res 15:371–382

55. Halina M, Ramesha S, Yarmob MA, Kamarudin RA (2007) Non-hydrothermal synthesis of mesoporous materials using sodium silicate from coal fly ash. Mater Chem Phys 101:344–351

56. Harvey RD, Cahill RA, Chou CL, Steele JD (1983) Mineral matter and trace elements in the Herrin and springfield coals, Illinois basin coal field. ISGS Contract/Grant Report 1983-4

57. Heidrich C, Feuerborn HJ, Weir A (2013) Coal combustion products: a global perspective. In: World of coal ash (WOCA) conference—April 22–25, 2013, Lexington

58. Hernandez-Exposito A, Chimenos JM, Fernandez AI, Font O, Querol X, Coca P, Garcia PF (2006) Ion flotation of germanium from fly ash aqueous leachates. Chem Eng J 118:69–75

59. Hill RO et al. (1981) Process of extraction of aluminium values from coal ash. US Patent No. 4243640

60. Hirajima T, Petrus HTBM, Oosako Y, Nonaka M, Sasaki K, Ando T (2010) Recovery of cenospheres from coal fly ash using a dry separation process: separation estimation and potential application. Int J Miner Process 95(1–4):18–24

61. Hoenderdal S, Espinoza LT, Marscheider-Weidemann F, Grus W (2013) Can a dysprosium shortage threaten green energy technologies? Energy 49:344–355

62. Hong K, Tokunaga S, Kajiuchi T (2000) Extraction of heavy metals from MSW incinerator fly ashes by chelating agents. J Hazard Mater B75:57–73

63. Huang K, Inoue K, Harada H, Kawakita H, Ohto K (2011) Leaching behavior of heavy metals with hydrochloric acid from fly ash generated in municipal waste incineration plants. Trans Nonferr Met Soc China 21:1422–1427

64. Huo P, Yan Y, Li S, Li H, Huang W (2009) Preparation and characterization of cobalt sulfophthalocyanine/$TiO_2$/fly-ash cenospheres photocatalyst and study on degradation activity under visible light. Appl Surf Sci 255:6914–6917

65. Hurst SL et al. (1978) Fly ash beneficiation process. US Patent No. 4121945

66. Hwang JY (1991) Wet Process for fly ash beneficiation. US Patent No. 5047145

67. Hwang JY (1993) Wet process for fly ash beneficiation. US Patent No. 5227047

68. Hwang JY, Sun X, Li Z (2002) Unburned carbon from fly ash for mercury adsorption: I. Separation and characterization of unburned carbon. J Miner Mater Charact Eng 1:39–60

69. Ilic M, Cheeseman C, Sollars C, Knight J (2003) Mineralogy of microstructure of sintered lignite coal fly ash. Fuel 82:331–336

70. Ishigaki T, Nakanishi A, Tateda M, Ike M, Fujita M (2005) Bioleaching of metal from municipal waste incineration fly ash

using a mixed culture of sulfur-oxidizing and iron-oxidizing bacteria. Chemosphere 60:1087–1094

71. Iwashita A, Sakaguchi Y, Nakajima T, Takanashi H, Ohki A, Kambara S (2005) Leaching characteristics of boron and selenium for various coal fly ashes. Fuel 84:479–485

72. Izquierdo M, Querol Z (2012) Leaching behaviour of elements from coal combustion fly ash: an overview. Int J Coal Geol 94:54–66

73. Jianguo Q et al. (2010) Process for recovery of silica followed by alumina from coal fly ash. US Patent No. 0119426A1

74. Jianguo Q et al. (2011) Process for recovery of silica followed by alumina from coal fly ash. US Patent No. 7871583B2

75. Jha VK, Matsuda M, Miyake M (2008) Resource recovery from coal fly ash waste: an overview study. J Ceram Soc Jpn 116:167–175

76. Joshi PB et al. (2015) Recovery of rare earth elements and compounds from coal ash. US patent No. 8968688B2

77. Joshi PB et al. (2013) Recovery of rare earth elements and compounds from coal ash. US patent No. 0287653A1

78. Karayigit AI, Onacak T, Gayer RA, Goldsmith S (2001) Mineralogy and geochemistry of feed coals and their combustion residues from the Cayirhan Power Plant, Ankara, Turkey. Appl Geochem 16:911–919

79. Ketris M, Yudovich Y (2009) Estimations of Clarkes for carbonaceous biolithes: world averages for trace element contents in black shales and coal. Int J Coal Geol 78:135–148

80. Krebs W, Brombacher C, Bosshard PP, Bachofen R, Brandl H (1997) Microbial recovery of metals from solids. FEMS Microbiol Rev 20:605–617

81. Kruger RA (1997) Fly ash beneficiation in South Africa: creating new opportunities in the market place. Fuel 76:777–778

82. Lisowyi B, Hitchcoc DC (1987) Beneficiation of gallium in fly ash. US Patent No. 4686031

83. Li L, Wu Y, Liu Y, Zhai Y (2011) Extraction of alumina from coal fly ash with sulfuric acid leaching method. Chin J Process Eng 11(2):254–258

84. Lisowyj B (1986) Method for extracting of iron, aluminium and titanium from coal ash. US Patent No. 4567026

85. Lisowyi B et al. (1986) Process for the recovery of gallium and germanium from coal fly ash. US Patent No. 4678647

86. Lisowyi B et al. (1987) Direct fuel-fired furnace arrangement for the recovery of gallium and germanium from coal fly ash. US Patent No. 4643110

87. Lisowyi B et al. (1987) Beneficiation of gallium in fly ash. US Patent No. 4686031

88. Lisowyi B et al. (1988) Condensation sand collection system. US Patent No. 4757770

89. Lisowyj B et al. (1988) Sand collection system. US Patent No. 4757772

90. Lakshman et al. (1990) Process for treatment of fly ash. US Patent No. 4966761

91. Loya MIM, Rawani AM (2014) A review: promising applications for utilization of fly ash. Int J Adv Technol Eng Sci 2(7):143–149

92. Lu S, Fang RL, Zhao H (2003) Study of recovery of highly pure super-fine powdered aluminum oxide from fly ash by way of lime sintered self powdering. Sci Study 1:15–17

93. Lukkassen D, Meidell A (2007) Advanced materials and structures and fabrication process. Book manuscript, Narvik University College, HiN

94. Masaaki N, Yoshihiro K, Setsuo T, Toru E, Iwao N (1996) Method for recovering vanadium from petroleum incineration ash. Mitsubishi Heavy Industries, Ltd. Eu Pat No. JP-A-H08325651

95. Maslov OD, Tserenpil S, Norov N, Gustova MV, Filippov MF, Belov AG, Altangerel M, Enhbat N (2010) Uranium recovery from coal ash dumps of Mogolia. Solid Fuel Chem 44:433–438

96. Matjie RH, Bunt JR, Heerden JHP (2005) Extraction of alumina from coal fly ash generated from a selected low rank bituminous South African coal. Miner Eng 18:299–310

97. Matyas B et al. (1987) Process for recovering rare metals from the combustion residue of coal by digestion. US Patent No. 4649031

98. Mayfield D, Lewis AS (2013) Environmental review of coal ash as a resource for rare earth and strategic elements. In: World of coal ash (WOCA) conference—April 22–25, 2013 in Lexington

99. McDowell WJ et al. (1981) Salt-soda sintering process for recovering aluminum from fly ash. US Patent No. 4254088

100. Meawad AS, Bojinova DY, Pelovski YG (2010) An overview of metals recovery from thermal power plant solid wastes. Waste Manag 30:2548–2559

101. Mehrotra AK, Behie LA, Bishnoi PR, Svrcek WY (1982a) High-temperature chlorination of coal ash in a fluidized bed. 1. Recovery of aluminium. Ind Eng Chem Process Des Dev 21:37–44

102. Mehrotra AK, Behie LA, Bishnoi PR, Svrcek WY (1982b) High-temperature chlorination of coal ash in a fluidized bed. 2. Recovery of iron, silicon, and titanium. Ind Eng Chem Process Des Dev 21:44–50

103. Mehrotra VP, Sastry KVS, Morey BW (1983) Review of oil agglomeration techniques for processing of fine coals. Int J Miner Process 11(3):175–201

104. Minter BE (2004) Method for recovering of trace elements from coal. US Patent No. 6827837B2

105. Mitner B (2005) Method for recovering trace elements from coal. US Patent No. 0056548A1

106. Mok WM, Willmes H, Wai CM (1984) Solvent extraction of molybdenum from biological samples and from coal fly ash for neutron activation analysis. Anal Chem 56:2623–2624

107. Moskaly RR (2004) Review of germanium processing worldwide. Miner Eng 17:393–402

108. Moskalyk RR, Alfantazi AM (2003) Processing of vanadium: a review. Miner Eng 16:793–805

109. Murtha MJ (1983) Process for the recovery of alumina from fly ash. US Patent No. 4397822A

110. Narukawa T, Riley K, French D, Chiba K (2007) Speciation of chromium in Australian fly ash. Talanta 73:178–184

111. National Research Council (US) (2006) Managing coal combustion residues in mines. National Academies, p 256

112. Navarro R, Guzman J, Saucedo I, Revilla J, Guibal E (2007) Vanadium recovery from oil fly ash by leaching, precipitation and solvent extraction processes. Waste Manag 27(3):425–438

113. Nayak N, Panda CR (2010) Aluminium extraction and leaching characteristics of Talcher thermal power station fly ash with sulphuric acid. Fuel 89(1):53–58

114. Ogata F, Tominaga H, Yabutani H, Taga A, Kawasaki N (2011) Recovery of molybdenum from coal by gibbsite. Toxicol Environ Chem 93:635–642

115. Okada T, Tojo Y, Tanaka N, Matsuto T (2007) Recovery of zinc and lead from fly ash from ash-melting and gasification-melting processes of MSW-comparison and applicability of chemical leaching methods. Waste Manag 27:69–80

116. Padil R, Sohn HY (1985) Sintering kinetics and alumina yield in lime-soda sinter process for alumina from coal wastes. Metallurg Mater Trans B 16:385–395

117. Pang J, LiQ Wang W, Xu X, Zhai J (2011) Preparation and characterization of electroless Ni–Co–P ternary alloy on fly ash cenospheres. Surf Coat Technol 205:4237–4242

118. Pennachetti J et al. (1973) Process for the treatment of fly ash. US Patent No. 3769054

119. Prakash S, Mohanty JK, Das B, Venugopal R (2001) Characterisation and removal of iron from fly ash of Talcher Area, Orissa, India. Miner Eng 14:123–126

120. Prasad B, Mondal KK (2009) Environmental impact of manganese due to its leaching from coal fly ash. J Environ Sci Eng 51(1):27–32

121. Qin S, Sun Y, Li Y, Wang J, Zhao C, Gao K (2015a) Coal deposits as promising alternative sources for gallium. Earth Sci Rev 150:95–101

122. Qin S, Zhao C, Li Y, Zhang Y (2015b) Review of coal as a promising source of lithium. Int J Oil Gas Coal Technol 9(2):215–229

123. Querol X, Fernández-Turiel J, López-Soler A (1995) Trace elements in coal and their behavior during combustion in a large power station. Fuel 74:331–343

124. Ramme BW et al. (2008) Separation of cenospheres from fly ash. US Patent No. 0190327A1

125. Ramme BW et al. (2011) Separation of cenospheres from fly ash. US Patent No. 8074804B2

126. Ramme BW et al. (2012) Separation of cenospheres from fly ash. US Patent No. 0050519A1

127. Ramya SS, Deshmukh VU, Khanderkar VJ, Padmakar C, SuriNaidu L, Mahore PK, Pujari PR, Panaskar D et al (2013) Assessment of impact of ash ponds on groundwater quality: a case study from Koradi in Central India. Environ Earth Sci 69:2437–2450

128. Rao RB, Chattopadhyay P, Banerjee GN (1999) Removal of iron from fly ash for ceramic and refractory applications. Magn Electr Sep 10:21–27

129. Reynolds RE et al. (1979) Process for Recovering aluminium and other metal values from fly ash. US Patent No. 4159310

130. Russ JJ et al. (1978) Process for recovering mineral values from fly ash. US Patent No. 4130627

131. Russ JJ et al. (1985) Method for separating and recovering silica, alumina and iron from fly ash, minerals ores and mine tailings. US Patent No. 4539187

132. Scharifker B et al. (2009) Process to recovery vanadium contained in acid solutions. US Patent No. 7498007B2

133. Seidel A, Zimmels Y, Armon R (2001) Mechanism of bioleaching of coal fly ash by *Thiobacillus thiooxidans*. Chem Eng J 83:123–130

134. Seredin VV (1996) Rare earth element-bearing coals from Russian Far east deposits. Int J Coal Geol 30:101–129

135. Seredin VV (2012) From coal science to metal production and environmental protection: a new story of success. Int J Coal Geol 90–91:1–3

136. Seredin V, Dai S, Sun Y, Chekryzhov IY (2013) Coal deposits as promising sources of rare metals for alternative power and energy-efficient technologies. Appl Geochem 31:1–11

137. Shabtai J et al. (1993) Extraction of metal oxides from coal fly ash by microorganisms and a new microorganism useful therefor. US Patent No. 5231018

138. Shiroto Y et al. (2003) Production of high purity vanadium compound from vanadium-containing carbonaceous residues. US Patent No. 6652819B2

139. Sierra-Alvarez R (2009) Removal of copper, chromium and arsenic from preservative-treated wood by chemical extraction-fungal bioleaching. Waste Manag 29:1885–1891

140. Sommerville R, Blissett R, Rowson N, Blackburn S (2013) Producing a synthetic zeolite from improved fly ash residue. Int J Miner Process 124:20–25

141. Soong Y, Schoffstall MR, Gray ML, Knoer JP, Champagne KJ, Jones RJ et al (2002) Dry beneficiation of high loss-on-ignition fly ash. Sep Purif Technol 26(2–3):177–184

142. Sparton Res Inc. (2007) First yellowcake produced from Xiaolongtang coal ash, China. http://www.spartonres.ca/pressreleases/PR2007Oct15.htm. Accessed 10 April 2014

143. Styron RW et al. (2000) Method of removing carbon from fly ash. US Patent No. 6068131

144. Tolhurst L (2015) Commercial recovery of metals from coal ash. In: World of coal ash (WOCA) conference in Nashville, May 5–7. http://www.flyash.info/

145. Torralvo FA, Fernández-pereira C (2011) Recovery of germanium from real fly ash leachates by ion-exchange extraction. Miner Eng 24:35–41

146. Toshiaki A, Akira S, Masutaka F (1998) Production of ammonium metavanadate. Eu Pat. No. JPH10251025

147. Ucurum M (2009) A study of removal of Pb heavy metal ions from aqueous solution using lignite and a new cheap adsorbent (lignite washing plant tailings). Fuel 88(8):1460–1465

148. US EPA (2005) Using coal fly ash in highway construction, a guide to benefits and impacts, EPA-530, K-05-002

149. US Environmental Protection Agency (1999) Technical background document for the report to congress on remaining wastes from fossil fuel combustion: waste characterization. EPA

530-R-99-010. Science Applications International Corporation, San Diego, CA

150. Vassilev SV, Vassileva CG, Karayigit AI, Bulut Y, Alastuey A, Querol X (2005) Phase-mineral and chemical composition of feed coals, bottom ashes and fly ashes at the Soma power station, Turkey. Int J Coal Geol 61:35–63

151. Vitolo S, Seggiani M, Filippi S, Brocchini C (2000) Recovery of vanadium from heavy oil and orimulsion fly ashes. Hydrometallurgy 57:141–149

152. Wang T, Wang J, Burken JG, Ban H, Ladwig K (2007) The leaching characteristics of selenium from coal fly ashes. J Environ Qual 36:1784–1792

153. Wang B, Li Q, Wang W, Li Y, Zhai J (2011) Preparation and characterization of $Fe^{3+}$ doped TiO2 on fly ash cenospheres for photocatalytic application. Appl Surf Sci 257(8):3473–3479

154. Wu C, Yu H, Zhang H (2012) Extraction of aluminium by pressure acid-leaching method from fly ash. Trans Nonferr Met Soc China 22:2282–2288

155. Xiang-yang C, Xin-zhe L, Qui-li Z, Hong-zhou MA, Hin Z (2010) Leaching vanadium by high concentration sulfuric acid from stone coal. Trans Nonferr Met Soc China 20:123–126

156. Xu M, Yan R, Zheng C, Qiao Y, Han J, Sheng C (2003) Status of trace element emission in a coal combustion process: a review. Fuel Process Technol 85:215–237

157. Xu X, Li Q, Cui H, Pang J, Sun L, An H et al (2011) Adsorption of fluoride from aqueous solution on magnesia-loaded fly ash cenospheres. Desalination 272(1–3):233–239

158. Yao ZT, Xia MS, Sarkar PK, Chen T (2014) A review of the alumina recovery from coal fly ash, with a focus in China. Fuel 120:74–85

159. Yao ZT, Ji XS, Sarker PK, Tang JH, Ge LQ, Xia MS, Xi YQ (2015) A comprehensive review on the applications of coal fly ash. Earth Sci Rev 141:105–121

160. Yavuz CT, Prakash A, Mayo JT, Colvin VL (2009) Magnetic separations: from steel plants to biotechnology. Chem Eng Sci 64(10):2510–2521

161. Zhang FS, Itoh H (2006) Extraction of metals from municipal solid waste incinerator fly ash by hydrothermal process. J Hazard Mater 136:663–670

162. Zhang Y, Talbott JL, Wiedenmann L, DeBarr J, Demir I (1999) Determination of germanium content in coal ash by wavelength-dispersive X-ray florescence spectroscopy and inductively coupled—mass spectrometry. JCPDS-International Center for diffraction data

163. Zyryanov VV, Petrov SA, Matvienko AA (2011) Characterization of spinel and magnetospheres of coal fly ashes collected in power plants in the former USSR. Fuel 90(2):486–492

# Chapter 4

# Demonstrating Fuel Flexibility in Commercial Operation*

**Fawad Khan*, Francis Lau**

*Synthesis Energy Systems Inc., Houston, USA*

## ABSTRACT

Synthesis Energy Systems Inc. (SES) enables its customers to advance gasification projects through commercial scale feedstock testing at its commercial scale coal to syngas facility in Zhao Zhuang City (ZZ Plant), Shandong Province, China. These tests help customers affirm individual project economic viability and reduce technology risk. Included in these test campaigns have been one and two-week long continuous operations at varying loads and on widely-varying feedstock such as Chinese Inner Mongolian lignite, Chinese sub-bituminous high ash (and high friable) coal, and Australian Queensland high ash sub-bituminous coal. Through these campaigns, SES has progressed one project into the detailed engineering and construction phase and one project into the feasibility study phase. SES believes that the ability to perform these commercial scale campaigns without interruption of operation is unique to SES. The ZZ Plant, which sells clean syngas to a neighboring methanol facility, has been in operation since early 2008 on a design coal of high ash middlings (washery wastes) from ROM bituminous coal (30% - 40% wt ash). SES has made significant improvements to the design, operation, and efficiency of the ZZ Plant and U-GAS technology that is licensed from Gas Technology Institute (GTI). Included in these is the Fines Management System (FMS), which has demonstrated overall carbon conversion of greater than 98%. The ZZ Plant consistently demonstrates fuel flexibility and greater than 98% availability, 98% carbon conversion, and 80% cold gas efficiency.

**Keywords:** Gasification; U-GAS; Fluid Bed Gasification; Low Rank Coal; Lignite, Biomass

## 1. INTRODUCTION

SES is a global energy and gasification technology company that provides products and solutions to the energy and chemicals industries. SES operates in regions where low-rank coal and biomass feedstocks can be profitably converted into high-value products through the application of SES' gasification technology. SES unlocks the value of these low-cost resources by providing a proprietary technology package whereby SES licenses its technology rights to third parties, delivers an engineered technology transfer package, and provides proprietary equipment components to customers who plan to own and operate projects. In addition, SES may partner with engineering, equipment and technology companies to provide its gasification technology package with an integrated modular product offering; invest in U-GAS (registered trademark of Gas Technology Institute) based projects either directly or through an investment partner; or acquire or partner with owners of low quality coal resources to create more value and opportunity for SES through the integration of its technology with the coal resources.

SES' proven gasification technology originated in the United States at the Gas Technology Institute (GTI) over 35 years ago. Since 1975, 11 pilot, demonstration, and commercial units have been designed and constructed. SES has demonstrated organizational technology development and operational skills that have led to an increase in retained proprietary know-how, trade secrets and technology capability. Furthermore, SES has a demonstrated track record of successfully implementing U-GAS with its first build-own-operate (BOO) project having over three years of successful operations and its second project on schedule.

## 2. SES' GASIFICATION PROCESS

SES' gasification process, based on U-GAS, is a singlestage,, fluidized-bed (bubbling-bed) technology for the production of low-to-medium heating value syngas from a wide array of coal and biomass feedstocks using oxygen, oxygen enriched air, or air in the gasification reaction. Within the bubbling bed, the feedstock reacts with steam and oxygen (either pure oxygen or oxygen in air). The reaction temperature is controlled to maintain high carbon conversion and non-slagging conditions for the ash. Carbon conversion and cold gas efficiencies of over 98% and 80%, respectively, have been repeatedly demonstrated at commercial scale on a wide range of feedstocks including low quality coals and lignite. Figure 1 is a typical block flow diagram of SES' gasification process base configuration.

The gasifier feed system consists of weigh bins, conveyors, transporters and lockhopper systems that supply the gasifier with coal or biomass at pressure. The feedstock is pressurized in a lockhopper system and metered in to the gasifier. Carbon dioxide or nitrogen is typically used as the feedstock transport gas medium into the gasifier reactor. The gasification agents (steam and oxygen) enter the gasifier through the grid, annulus and the center jet at the bottom of the gasifier as illustrated in Figure 2.

The gasification system consists of the gasifier, fines removal, ash discharge, and ash cooling systems. Within the fluidized bed, the feedstock reacts with steam and oxygen (either pure oxygen or oxygen in air). The process accomplishes four important functions in a singlestage bubbling-bed gasifier: it decakes, devolatilizes, and gasifies the feedstock, and if necessary, agglomerates and separates ash from the reacting coal. At the specified operating conditions, coal is gasified to produce a synthesis gas ("syngas") product consisting of hydrogen, carbon monoxide, carbon dioxide, water vapor, and methane, in addition to small amounts of ammonia, hydrogen sulfide, and other trace impurities.

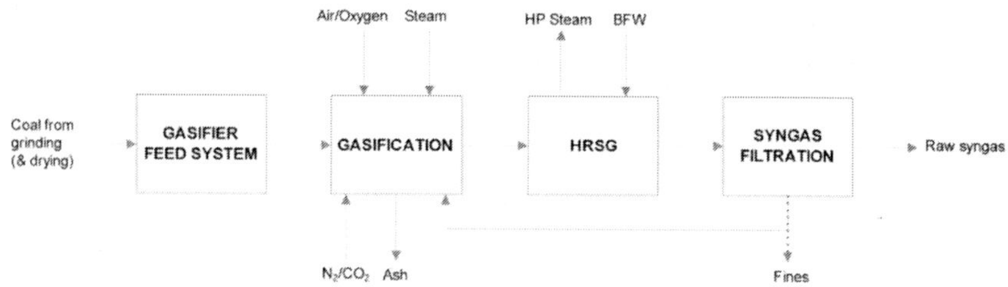


**Figure 1.** Typical SES' gasification process configuration.

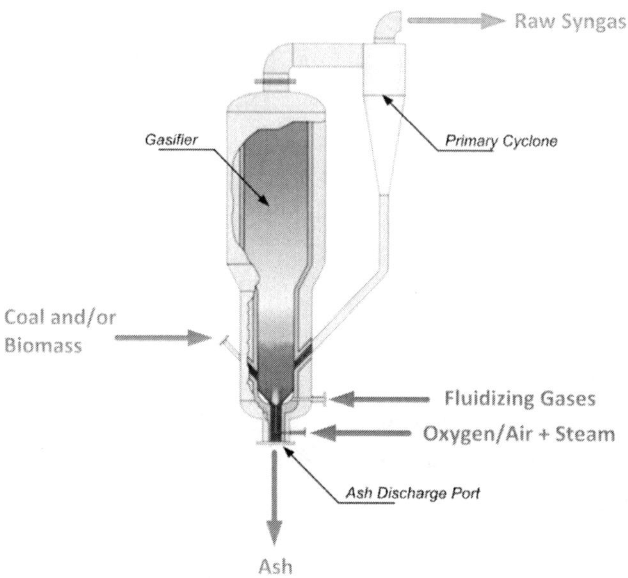


**Figure 2.** Gasifier configuration.

The bottom ash flow rate through the ash discharge port is adjusted to obtain adequate ash discharge from the gasifier. The temperature and pressure of the coarse ash from the gasifier is reduced as ash flows through the ash classifier and bottom ash handling system. The bottom ash, upon leaving the ash classifier is, cooled and removed from the plant via an ash cooler, lockhopper system, and screw coolers before being transported outside by belt conveyors for truck unloading.

The primary fines recovery and recycle system consists of two cyclones in series, designated as the primary and secondary cyclones, which recover nearly all fines from the syngas stream leaving the gasifier. The primary cyclone is designed to efficiently recover entrained fines from the syngas with a relatively small pressure drop. The fines collected in the primary and secondary cyclones are returned to the gasifier by means of a dip-leg.

A heat recovery steam generator (HRSG) serves to increase the energy efficiency of the gasification system. Steam generated from the HRSG that is not used by the gasification process is exported to meet other plant needs including power generation. Cooled syngas leaving the HRSG then enters the downstream high-efficiency cyclone and filter system.

The collected fines from the high-efficiency cyclone and filter system are routed to a fines silo through a lockhopper system, where they are collected in the baghouse and returned to the gasifier for further conversion. This system is the Fines Management System (FMS), recently incorporated into SES' U-GAS design for maximizing carbon conversion, and has been proven at commercial scale.

# 3. MAJOR PROCESS IMPROVEMENT—SES' FINES MANAGEMENT SYSTEM

Prior to 2010, the best-in-class carbon conversion for fluid bed gasification technologies was 97% [1]. In October 2010, SES implemented and successfully demonstrated its FMS at the ZZ Plant. Through its FMS, SES has been able to successfully recycle and gasify waste fines generated from the gasification process. The result has enabled SES' gasification process to achieve high carbon conversions (greater than 98%) and cold gas efficiencies (greater than 80%) on low quality feedstocks that may not be feasible to gasify in commercial slagging gasifiers. Furthermore, the FMS eliminates many challenges associated with fines (i.e. fly ash) disposal.

Figure 3 illustrates simulated gasifier performance on sub-bituminous coal for SES' gasification process with and without the FMS. Simulated gasfier performances on Illinois #6 coal from slagging dry and slurry fed entrained and fixed bed gasification processes have been provided as a reference.

The carbon conversion and cold gas efficiency in Figure 3 are calculated values. The Slagging wet and dry feed gasification calculations are based on Department of Energy's IGGC performance study [2]. The Fixed bed gasification calculations are based on Department of En ergy's SNG performance study [3]. The Cold Gas Efficiency is defined as follows:

$$\text{Cold Gas Efficiency (CGE)} = \frac{\text{Effective Syngas Output } \left(H_2 + CO + CH_4\right)\left[\dfrac{\text{mmbtu}}{\text{hr}}\right]}{\text{Coal Input }\left[\dfrac{\text{mmbtu}}{\text{hr}}\right]}$$

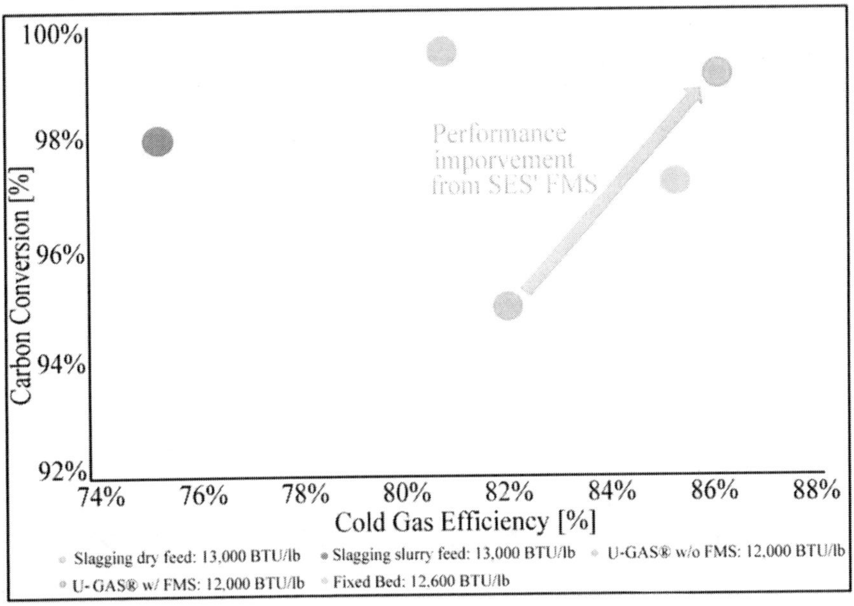


**Figure 3.** Comparative assessment of SES FMS system[2].

# 4. ZZ COMMERCIAL OPERATIONS

In January 2008, SES completed construction of its ZZ Plant in Zao Zhuang City, China. The ZZ Plant is a 96% SES owned joint venture with Shandong Hai Hua Coal & Chemical Company Ltd. ("Hai Hua"). The ZZ plant consists of two gasification trains designed to process 400 metric tons per day of coal middlings with up to 40 wt-% ash. Clean syngas is sold over the fence to the JV partner, Hai Hua, for methanol production. During the first year of commercial operations, the plant achieved over 98% availability of single gasifier capacity using low-value high-ash coal and coal washings. Moreover, the ZZ plant has achieved less than 1kg of coal per normal cubic meter (NCM) of clean syngas (primarily carbon monoxide, hydrogen, and methane) and less than 0.4 NCM of oxygen per NCM of clean syngas. Figure 4 is a process block flow diagram representation of the ZZ Plant configuration.

The ZZ Plant continuously operates at 90% carbon dioxide ($CO_2$) capture from syngas through pressure swing absorption. Part of this $CO_2$ is recycled back to the gasifier and the balance is directed to the flare.

# 5. FUEL FLEXIBILITY

A key advantage of SES' gasification technology is fuel flexibility. Typically, fluid bed gasification technologies require coal with high ash fusion temperatures (>1830˚F) to avoid ash agglomeration [4]. However, the U-GAS process has features in its design that allow the technology to manage coals with ash that may agglomerate. The technology has been proven on many ranks of coal (low rank, high

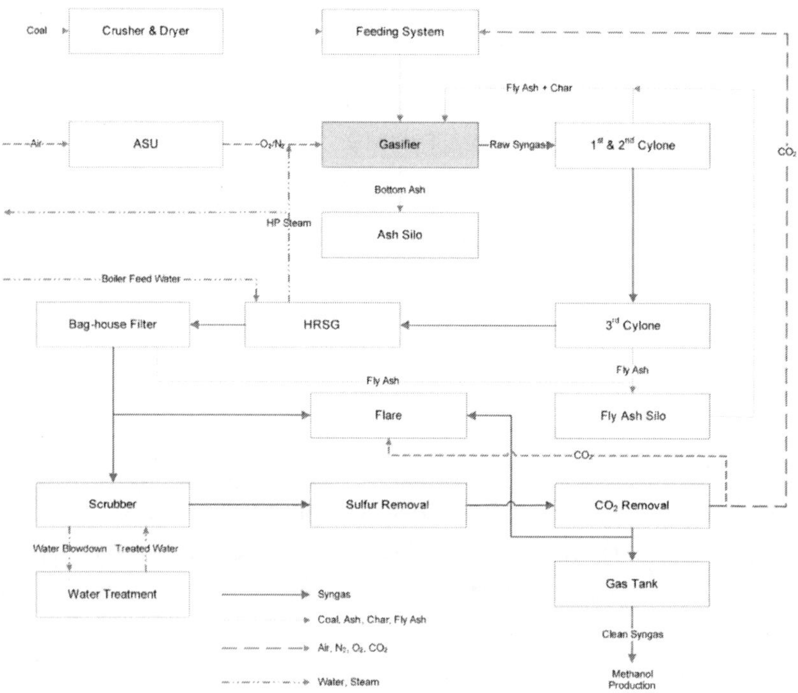


**Figure 4.** Zao Zhuang Plant block flow diagram.

moisture, and high ash coals including Indian coal), coal waste products, and biomass feedstocks. Table 1 and Table 2 represents SES' bench, pilot and commercial scale feedstock test experience to date. This information should not be interpreted as technical limitations to SES' gasification process. SES evaluates commercial and technical viability of feedstocks on a case by case basis. At the ZZ Plant, SES continues to prove feedstock flexibility by on-stream switching of feedstocks without interrupting syngas supply to Hai Hua for its methanol production.

In 2004 and 2005, Nexant conducted a successful evaluation of the U-GAS technology for USDAID/India in association with the National Thermal Power Company of India and USDOE/NETL. The evaluation included testing of Indian coal with 34 - 41 wt% ash from the North Karanpura coal fields. The ROM and washed Indian coals were successfully gasified under air-blown operating conditions in GTI's pilot scale U-GAS reactor at feed rates up 11 ston/day and pressures up to 304 psi(g) [5].

Since the commercial operation of its ZZ Plant in 2008, SES has successfully conducted test campaigns for third parties on off-design coals including high ash sub-bituminous coal, Inner Mongolian Lignite and high ash Australian brown coal. Operating flexibility is critical to a project's success where there is uncertainty in fuel availability and pricing. Figure 5 below provides a simplified operating envelope (fuel preference) when presented with a high moisture sub-bituminous and high ash coal.

The colors illustrate the pay-off opportunity by operating on the respective coal at the specified price combination. Traditional net present value approaches fail to capture such inherent value in operating flexibility that SES' technologies provide.

**Table 1.** Gasification feedstock experience.

| Coalk | Coal Source |
|---|---|
| **Bituminous Coals** | ▪ Western Kentucky No. 9, washed & Run-of-Mine (ROM)<br>▪ Western Kentucky No. 9 and 11, Camp<br>▪ Illinois No. 6, Peabody No. 10 and Crown III<br>▪ Pittsburgh No. 8, Champion and Ireland<br>▪ Australian, Bayswater No. 2, Sydney Basin<br>▪ Polish, Silesia<br>▪ French, Merlebach—ROM<br>▪ Utah - ROM<br>▪ Colombian<br>▪ Indian, North Karanpura, washed and ROM<br>▪ Shandong—High ash, ROM, and coal washing middlings |
| **Biomass** | ▪ Finnish waste wood and pulp mill waste<br>▪ Danish Willow<br>▪ Danish Straw<br>▪ Pelletized alfalfa stems<br>▪ Pelletized waste wood<br>▪ Bagasse<br>▪ Rice straw<br>▪ Chicken litter |

| | |
|---|---|
| **Sub-Bituminous and Lignite** | ▪ Chinese, Shen Fu Sub-Bituminous<br>▪ Henan Yima Sub-Bituminous<br>▪ Montana Rosebud, Colstrip<br>▪ Wyoming, Big Horn<br>▪ North Dakota lignite, Freedom<br>▪ Saskatchewan lignite<br>▪ Inner Mongolia lignite |
| **Coke, Char, Peat, & Waste** | ▪ Metallurgical Coke, U.S., China, Poland<br>▪ Western Kentucky No. 9 coal char<br>▪ Illinois No. 6 coal char<br>▪ Finnish Peat, Viidansuo and Savaloneva<br>▪ Automobile Shredder Residue<br>▪ Oil shale<br>▪ RDF and Auto shredder Residue |

**Table 2.** Tested feedstock property ranges.

| Feedstock Property | Tested Range |
|---|---|
| Moisture Content, wt% | 1 - 41 |
| Volatile Matter, wt% | 3 - 69 |
| Fixed Carbon, wt% | 6 - 83 |
| Sulfur, wt% | 0.2 - 4.6 |
| Free Swelling Index | 0 - 8 |
| Ash Content, wt% | <1% - 55% |
| Ash Softening—$T_1$, °F | 19000 - 26600+ |
| Heating Value, HHV, BTU/lb | 5490 - 13860 |

# 6. PHASED APPROACH

Feedstock characterization and testing campaigns fit well with and can accelerate a staged approach to project development. Each stage is an option that is a right but not an obligation to take action in the future. While commercial gasification technology platforms are well known to provide the option for carbon capture, SES' gasification technology adds additional value by providing optionality through fuel flexibility, as discussed above, and staging of project execution. SES is in the unique position to perform the following feedstock validation testing for its customers to provide an optimum plant design, plan for technical and commercial risk mitigation, and customer assurance on gasification performance for the chosen feedstocks. The four main tests are:

1) Feedstock properties evaluation;

2) Bench scale gasification;

3) Pilot scale gasification;

4) Commercial scale gasification.

As illustrated in Figure 6, each stage provides an opportunity for project viability assessment, and allows customers an off-ramp to project development spending if the project does not appear to be viable at that particular stage. Proving feedstock applicability in a commercial scale allows customers to accelerate the development of a project by removing or reducing feedstock and gasification technology risk. The proactive approach results in a higher net present value ("Active NPV") that shall exceed net present value from the passive approach to project development ("Passive NPV") due to higher risks associated with the latter.

SES has demonstrated this risk reducing-staged approach to project execution through its joint venture with Yima Coal Industry Group ("Yima"), which is SES' second project that is under construction in

Henan Province, China. SES is a 25% owner of this joint venture. Following preliminary coal characterization and testing, SES leveraged its commercial testing capabilities at the ZZ Plant, which shifted from the plant's design coal (40 wt% ash bituminous coal washing by-product) to a subbituminous coal with 33 - 37 wt% ash that was provided by Yima. During the test, the plant achieved high carbon conversion rates and full load throughput. This test along with other coal testing campaigns provided a clear demonstration of the flexibility of SES' technology in gasifying many ranks of coal efficiently. Of equal importance is the fact that Yima owns and operates an entrained flow gasification plant and one of Asia's largest fixed bed gasification plants but selected SES' gasification process over these alternative gasification techno logy platforms for Yima's new coal-to-chemical park.

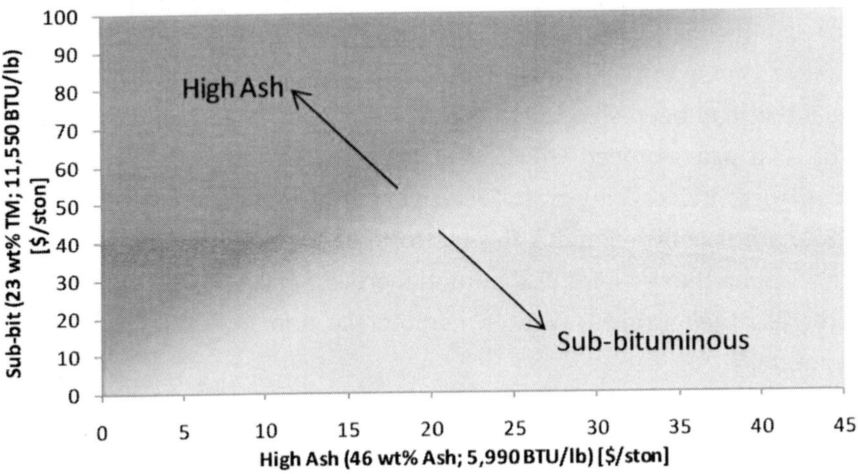


**Figure 5.** Example U-GAS fuel selection flexibility.

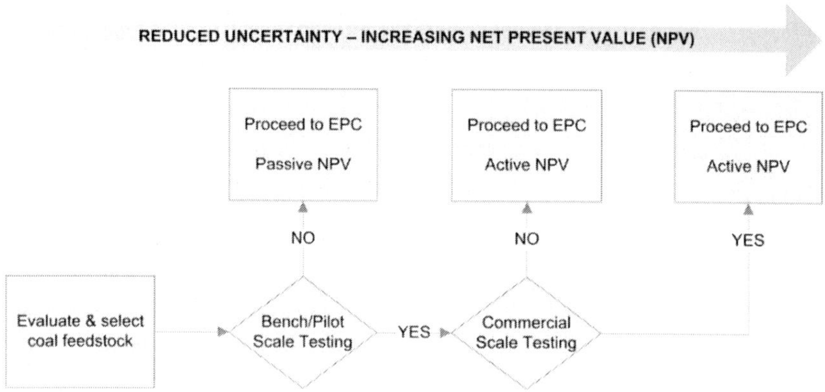


**Figure 6.** Options valuation through project staging.

SES' ZZ Plant continues to enable customers to meet their project needs through coal campaigns on feedstock imported by potential licensees for commercial scale testing. In December 2010, SES worked with Ambre Energy to test almost 3300 short tons of their Australian brown coal that had been shipped to the ZZ Plant from Queensland, Australia. This test again proved U-GAS efficiency and effectiveness on difficult feedstocks by reaching over 82% cold gas efficiency and greater than 98% carbon conversion. It's this platform of commercial scale feedstock gasification that enables SES' potential project partners and licensees to progress their projects with a commercial and bankable technology solution.

## 7. CONCLUSION

SES has made significant improvements to the design, operation, and efficiency of the U-GAS technology and the ZZ Plant. Included in these improvements is the FMS. Moreover, the ZZ Plant consistently demonstrates operating flexibility and greater than 98% availability, 98% carbon conversion, and 80% cold gas efficiency. SES provides its customers with real options in the form of operating flexibility and phased project development. Both have been demonstrated by SES through existing operations at the ZZ Plant and the execution of SES' Yima joint venture project. Growing interest by third parties in SES' gasification solutions, including the successful commercial testing and feasibility study of Ambre Energy's high ash Queensland coal, is further testament to the advantages presented by SES.

## REFERENCES


1. C. Higman and M. van der Burgt, "Gasification," Gulf Professional Publishing, 2008.

2. M. C. Woods, P. J. Capicotto, J. L. Haslbeck, N. J. Kuehn, M. Matuszewski, L. L. Pinkerton, et al., "Cost and Performance Baseline for Fossil Energy Plants DOE/NETL- 2007/1281," National Energy Technology Laboratory, 2007.

3. L. M. Bartone and J. White, "Industrial Size Gasification for Syngas, Substitute Natural Gas and Power Production DOE/NETL-401/040607," National Energy Technology Laboratory, 2007.

4. A.-G. Collot, "Matching gasification technologies to coal properties," International Journal of Coal Geology, 2006, pp. 191-212.

5. B. Bryan, R. Laurens, A. Kramer, J. Hudziak, F. Lau, T.-P. Chen, et al., "Flex-Fuel Test Facility: Evaluation of Fluidized-Bed Gasification for IGCC Power Production with High-Ash Indian Coal," International Pittsburgh Coal Conference, 2005.

# Chapter 5

# Coal Washing Scenario in India and Future Prospects

**R. Venugopal[1] • J. P. Patel[2] • C. Bhar[3]**

[1] *Dept. of Fuel and Mineral Engineering, Indian School of Mines, Dhanbad, Jharkhand 826004, India*

[2] *Washery Division, Bharat Coking Coal Limited, Saraidhela, Dhanbad, Jharkhand 828127, India*

[3] *Dept. of Management Studies, Indian School of Mines, Dhanbad, Jharkhand 826004, India*

## ABSTRACT

Coal Washing Exploration in India dates back to 1900s; though, first coking coal washeries in India were installed after independence. At present, most of the coking coal washeries are owned by Public Sector Companies; whereas, most of the non-coking coal washeries are owned by Private Sector. Even after six decades of coal washing practices, there has not been significant development in the coal washing intelligentsia. Indian Coal Washing industry is still dependent on imported equipment, which has been designed to treat coal that is significantly different from Indian coal of drift origin. In this paper, authors have ventured into evolution of Indian Coal Washing Industry (with a focus on coking coal washing sector), its present condition and future prospect for growth. The paper emphasizes need for developing indigenous solutions to industrial challenges and highlights importance of increased coordination among academia-research institutions and coal industry.

**Keywords :** Coking coal washingIndian coal industryHistory of coal washing Future prospects

## *1* EVOLUTION OF COAL WASHING IN INDIA

After setting up Steel Plants in India (in 1907), Tata Iron and Steel Company (TISCO) started explorations to meet the demand of coking coal for their steel plants; and, first breakthrough was achieved by Mr. A. Farquhar during his washability studies of Jharia and Raniganj coal during 1918–1926. After ascertaining economic viability, in 1946, TISCO decided to set up Coal Washeries in West Bokaro (1951) and Jamadoba (1952) (Ghosh 1964; Sen 2015). Chance Cone separator (a dense media separator utilizing sand as media) was the first washing equipment installed in Jamadoba Coal Preparation Plant (JCPP) for treatment of Jamadoba and Bhowra Coal. Subsequently, Lodna Coal Company set up Feldspar Jig—based very simple Coal Washery in 1955. (Ghosh 1964; Sen 2015; Singh and Guha 1965; Sinha 1988).

Hindustan Steel Limited (HSL) was set up in 1954, for expansion of Steel Industry in India. Similarly, National Coal Development Corporation (NCDC) was founded for planned growth of coal mining industry in India. To meet demand of coking coal for HSL Steel Plants, Central Coal Washeries in the vicinity of existing Marshaling yards of Kargali, Durgapur, Dugda, Bhojudih and Patherdih were envisaged; and Kargali Washery was commissioned in 1958 by NCDC as first public sector washery in India. Subsequently, Durgapur, Dugda, Bhojudih and Patherdih Coking Coal Washeries were commissioned during the III and IV Five Year Plans. (CCL 2015; CIL 2015; Ghosh 1964; CCL 2015; AICC 1965; Singh and Guha 1965; Sarkar 1972; Das 1975; Agrawal 1976; CEMPDIL 2016; Konar et al. 1998).

Following the footprints, Indian Iron and Steel Company (IISCO), also set up Chasnalla Washery in 1968 (Sarkar 1972; Agrawal 1976; Konar et al. 1998; Rai and Tiwari 2009; Sen 2015). Later on, NCDC installed Kathara and Swang Washeries in 1969 and 1970. After nationalization of coal industry in India, ownership of NCDC Washeries was transferred to Central Coalfields Limited (CCL), (CEMPDIL 2016).

After a decade, for administrative reasons, ownership of SAIL Washeries was transferred to Bharat Coking Coal Limited (BCCL). To cater for the growing demand of clean coal, Coal India commissioned seven more coking coal Washeries (Sudamdih, Moonidih, Nandan, Barora, Mohuda, Kedla and Madhuband) during 1980–2000 (Ministry of Coal 2014; Sen 2015; Sinha 1988; Saxena 2015; NSO 2015). The most recent ones are Kedla and Madhuband Washeries of CCL and BCCL commissioned in 1997 and 1998 respectively.

Over the period, quality of coal has deteriorated due to exhaustion of superior grade coal in upper seams of BCCL and CCL Coal Belt (Sarkar et al. 2015). However, there has not been significant development in coal washing circuits, as the most recent washeries (Kedla and Madhuband) are Batac Jig-based washeries in contrast with the earliest coal washeries of HSL, which comprised of Heavy Media Separators (Rai and Tiwari 2009).

The demand of coal increased with the growth of Steel Industry. Keeping this in view, Coal India has planned to install six new coking coal washeries in BCCL with design capacity of 18.6 MTPY. These washeries will be installed and run on BOM scheme (Build Operate and Maintain). Three of these washeries are under construction. Five stage of Washing is being introduced in India with these washeries (Sarkar et al. 2015; Saxena 2015).

## 2 GROWING IMPORTANCE OF WASHING

India has proven coal reserves of total 125.91 billion tons, out of which only 18.4 billion tons is that of coking coal as shown in Table 1 (Source: Ministry of Coal).

**Table 1.** Geological reserves of coal in India as on 01.04.2014 (Quantity in MT)

| Type of coal | Proved | Indicated | Inferred | Total |
|---|---|---|---|---|
| (A) Coking | | | | |
| Prime coking | 4614 | 699 | 0 | 5313 |
| Medium coking | 13,303 | 11,867 | 1879 | 27,049 |
| Semi-coking | 482 | 1004 | 222 | 1708 |
| Sub-total coking | 18,400 | 13,569 | 2101 | 34,070 |
| (B) Non-coking | 106,916 | 128,838 | 30,249 | 266,002 |
| (C) Tertiary coal | 594 | 99 | 799 | 1493 |
| Total | 125,909 | 142,507 | 33,149 | 301,565 |

Over the years, prime-coking coal reserves in the Jharia Coal Field have been exploited from the upper seams, leaving behind lower seams of inferior grade coal for further extraction.

Grade wise distribution of various seams of coal in BCCL (the only producer of prime-coking coal in India) is indicated in Table 2 (Sarkar et al. 2015).

**Table 2.** Grade wise distributions of coal seams/reserves in Bharat Coking Coal Limited

| Demarcation | Coal seam | Thickness (m) | Grade of coal |
|---|---|---|---|
| Upper horizon | XVIII A/B/C/D | 5.9 | ST-II and W-I |
| | XVII | 4.45 | coking |
| | XVI A/B/C/D/E | 10.58 | |
| | XV | 2.54 | W-I and W-II |
| | XIV | 0.56 | coking |
| | XIII | 3.94 | |
| | XII | 6.25 | W-II coking |
| | XI/XII | 6.83 | |
| | IX/X, X | 10.57 | |
| Lower horizon | VIII A,B | 7.84 | NLW |
| | VIII | 2.62 | W-IV |
| | V/VI/VII | 20.84 | |
| | IV/(T)/IV (B) | 8.4 | Non-coking |

Unlike metallurgical grade coal (ST-I & ST-II), inferior grade coal cannot be directly blended with imported coal for coke making. Increasing demand of growing steel industry and diminishing production of metallurgical grade coal has resulted in increasing import burden on Indian Steel Industry to meet its demand for quality coking coal (Ministry of Coal 2014; CIL 2015). Trend of gross national demand for coking coal, domestic production and import during last six years has been shown in Fig. 1 (Ministry of Coal 2014; NSO 2015).

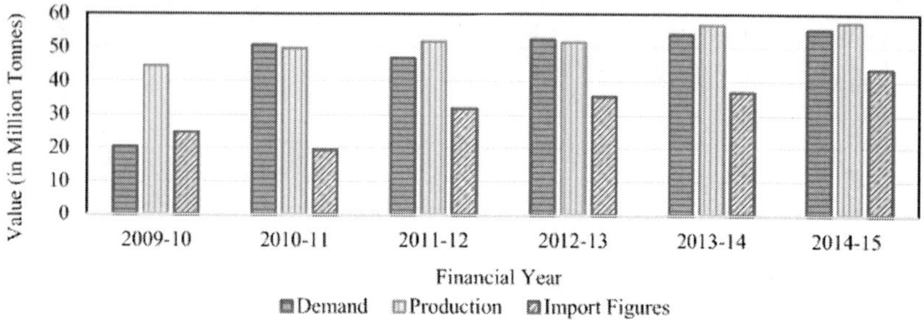


**Figure 1.** Trend of gross national demand, annual production and import of coking coal in India

Statistically, domestic production is enough to meet demand of coal for the Indian Steel Industry; still, import of coking coal has only been increasing. Inferior quality of Indian coking coal, and insufficient washing capacity are the principal reasons behind this huge demand–supply gap. Approx. 79 % of the total national demand is met by imported coal. Quality of Coking Coal produced during 2014–2015 is indicated in Table 3. (Ministry of Coal 2014; NSO 2015).

**Table 3.**Grade of coking coal produced in India

| Grade of coal | ST-I | ST-II | Semi coking | W-I | W-II | W-III | W-IV |
|---|---|---|---|---|---|---|---|
| Production in MT | 0.50 | 0.456 | 0.13 | 0.115 | 2.228 | 12.499 | 41.973 |

It is evident that, about 95 % of total coal produced in India is of Washery Grade -III and IV. As per Coal Statistic Report, only 24 % of the total coking coal produced in India has been utilized in metallurgical sector; and remaining 76 % has been diverted to non-metallurgical sector. The figures are even dismal, if only public sector companies are considered, with a meager 14.4 % of the total coking coal utilization in metallurgical sector. The situation is alarmingly critical as we are not only losing valuable coking coal reserves to non-metallurgical sectors, it's costing Indian foreign currency reserves to the tune of US $5.2 billion in import of coking coal (Ministry of Coal 2014; CIL 2015).

## 3 EXISTING SCENARIO OF COKING COAL WASHING IN INDIA

So far, 25 coking coal Washeries have been installed and commissioned in India out of which only 19 are in operation at present with a total washing capacity of 33.17 MTY; out of these Washeries, 15 are operating in Public Sector. (Sen 2015; Saxena 2015; NSO 2015 ; Konar et al. 1998; AICC 1965; Sinha 1988; Rai and Tiwari 2009). It is evident from Table 3 and Table 4 that the existing capacity of Washeries are not sufficient to handle total amount of coking coal produced in the country, and remaining quantity of coal is bound to get diverted to non-metallurgical sector. Moreover, it is also to be noted that only 13.79 MT of Coking Coal is being channelized to Metallurgical Sector, which amounts to only 41.6 % of the total washing capacity (Ministry of Coal 2014). This poor capacity utilization can be attributed to old plant machinery and obsolete washing circuit in most of the existing washeries, which have outlived their-designed life. Recurring losses in the coking coal washeries is another principal deterrent in capacity utilization (Sarkar et al. 2015). On further investigation, it was revealed that these washeries are operating at very poor yield; and in many of the public sector washeries fine coal circuit is not in operation (Sarkar et al. 2015; Alam and Patel 2012). Thus, actual utilization of coking coal in the metallurgical sector is even less, not more than 10.0 %. The authors are carrying out studies on Coking Coal Washeries of Bharat Coking Coal Limited to help improve their performance and ultimately maximize the optimum utilization of coking coal in the metallurgical sector.

**Table 4.**Existing Coking Coal Washeries in India and Washing Technologies

| S. no. | Name of washery | Year of commissioning | Design capacity (MTPY) | Washing technology | | |
|---|---|---|---|---|---|---|
| | | | | Coarse coal | Small coal | Fine coal |
| 1 | Jamadoba | 1952 | 0.9 | – | −13 + 0.5 mm (HM Cyclone) | −0.5 mm (Flotation) |
| 2 | Kargali | 1958 | 2.78 | −300 + 30 mm (Coarse Coal Jig) | – | – |
| 3 | Durgapur[b] | 1960 | 1.5 | −80 + 20 mm (H M Bath) | −20 + 1.0 mm (Small Coal Jig) | −1 + 0.25 mm (Spiral) −0.25 (Flotation) |
| 4 | Bhojudih | 1962 | 1.7 | −75 + 25 mm (HM Bath) | −25 + 0.5 mm (Batac Jig) | −0.5 mm (Flotation[a]) |
| 5 | Dugda-II | 1968 | 1.8 | – | −13 + 0.5 mm (HM Cyclone) | −0.5 mm (Flotation[a]) |
| 6 | DCOP | 1968 | 1.35 | Not Available | | |
| 7 | Chasnalla | 1968 | 1.5 | −80 + 20 mm (H M Bath) | −20 + 0.5 mm (Small Coal Jig) | −0.5 + 0.15 mm (Spiral) −0.15 (Clean Coal) |
| 8 | Kathara | 1969 | 3.0 | −80 + 13 mm (Drewboy Bath) | −13 + 0.5 mm (HM Cyclone) | −0.5 mm (Flotation[a]) |
| 9 | Swang | 1970 | 0.75 | −80 + 20 mm (Deshaler Jig) | −20 + 0.5 mm (HM Cyclone) | −0.5 mm (WO Cyclone[a]) |
| 10 | Sudamdih | 1981 | 1.6 | – | −37 + 0.5 mm (HM Cyclone) | −0.5 mm (Flotation[a]) |
| 11 | West Bokaro-II | 1982 | 1.8 | – | −13 + 0.5 mm (HM Cyclone) | −0.5 mm (Flotation) |
| 12 | Moonidih | 1983 | 1.6 | – | −30 + 0.5 mm (HM Cyclone) | −0.5 mm (WO Cyclone[a]) |
| 13 | Nandan | 1984 | 1.2 | −75 + 13 mm (Coarse Coal Jig) | −13 + 0.5 mm (Fine Coal Jig) | −0.5 mm (Flotation[a]) |
| 14 | Rajrappa | 1988 | 3.0 | −80 + 16 mm (Batac Jig) | −16 + 0.5 mm (Fine Coal Jig) | −0.5 mm (Flotation[a]) |
| 15 | Mohuda | 1990 | 0.63 | – | −25 + 0.5 mm (HM Cyclone) | −0.5 mm (Flotation[a]) |
| 16 | Bhelatand | 1994 | 0.86 | – | −13 + 0.5 mm (HM Cyclone) | −0.5 mm (Flotation) |
| 17 | West Bokaro-III | 1994 | 2.1 | – | −13 + 0.5 mm (HM Cyclone) | −0.5 mm (Flotation) |
| 18 | Kedla | 1997 | 2.6 | −80 + 15 mm (Batac Jig) | −15 + 0.5 mm (Fine Coal Jig) | −0.5 mm (Flotation[a]) |
| 19 | Madhuband | 1998 | 2.5 | – | −13 + 0.5 mm (Batac Jig) | −0.5 mm (Flotation[a]) |

[a] Fine coal circuit is not in operation
[b] Details pertaining to operation of fine coal circuit not available

Coal Washing Technology has evolved globally. Some countries have almost frozen their washing technology for a particular coal seam. This is not the case with Indian coal industry. The Washing Technology developed elsewhere, have been slightly customized to face the challenges of difficult washing characteristics of Indian Coal. As such, no washing equipment has been designed specifically to treat Indian Coal with difficult washing characteristics, which is not encountered in case of carboniferous coal. Details of washing schemes in existing coking coal washeries of India has been compiled in Table 4 (Sen 2015; Ministry of Coal 2014; Chakraborty et al. 2015; Sarkar et al. 2015; CCL 2012; Rai and Tiwari 2009). Heavy Media Cyclone is the most widely used separator in these washeries. Presence of high proportion of near-gravity-material (NGM) mandates cyclone washing (Rai and Tiwari 2009). Misplacement of coal in such washing schemes can be further improved with the aid of automation and control. However, most of the public sector washeries are yet to integrate automation and control in their existing circuits.

## *4* CHARACTERISTIC OF INDIAN COAL

Indian coals of drift origin (Gondwana Coal) can be characterized by high ash content with impurities disseminated in the coal bed. Very thin bands (5 mm–3 cm) of impurities are often encountered in between coal bands. Such coal has high near gravity material (NGM > 30 %) content. Liberation size of coal is much smaller; and significant improvement in yield is observed when coal is crushed down to as low as −6 mm to −3 mm and −1 mm. However, operating a coal washery with such small coal size is not feasible under present techno-economics. This makes Indian Coal difficult to wash, resulting in poor yield and low organic efficiency.

Indian industries find Indian Coal inferior to the imported coal mainly because of high ash content. However, Indian Coal can also be characterized by low sulfur (0.2 %–0.7 %), low iron, low chlorine, low toxic/rare earth elements content and macerals rich coal with high ash fusion temperature, and refractory nature of silica and alumina rich ash. It offers great potential for further investigation into coal washing technology dedicated for Indian Coal.

# 5 FUTURE OF COAL WASHING INDUSTRY

Government of India planned to expand coal-washing capacity in India by 13th Five Year Plan. Installation of 20 Coal Washeries in public sector with total installed capacity of 111.10 MT was envisaged. At present, CIL has taken up installation of 15 coal washeries on build, operate and maintain (BOM) Scheme.

Out of these, six are coking coal washeries, under the ownership of Bharat Coking Coal Limited with total design capacity of 18.6 MTY; and the remaining nine Washeries are Non-Coking Coal Washeries with total designed capacity of 74.5 MTY under the ownership of CCL (03 No—19.5 MTY), SECL (02 No—15 MTY) and MCL (04 No—40 MTY) subsidiaries of Coal India Limited (Saxena 2015; Sarkar et al. 2015).

The six coking coal washeries of Bharat Coking Coal Limited will be equipped with advanced automation and control circuits. Three of these washeries are under construction with the remaining under various stages of finalization. Typical cyclone—Cyclone, Jig—Cyclone combination has been used. deshaling cyclone and teeter bed separator (TBS) are being used for the first time in India.

A brief description of washing technology proposed in these washeries is presented in Table 5 (Sarkar et al. 2015; Sapru 2015).

**Table 5.**Upcoming coking coal washeries in coal India limited

| Name of washery | Design capacity | Washing technology | | |
|---|---|---|---|---|
| | | Coarse coal | Small coal | Fine coal |
| Madhuband | 5.0 MTY | −75 + 13 mm (Deshaling cyclone) | −13 + 1 mm (HM Cyclone) | −1 + 0.25 mm (TBS) −0.25 mm (Flotation) |
| Patherdih | 5.0 MTY | −75 + 13 mm (Jig) | −13 + 0.5 mm (HM Cyclone) | −0.5 mm (Flotation) |
| Dahibari | 1.6 MTY | −50 +13 mm (Deshaling Jig) | −13 + 0.5 mm (HM Cyclone) | −0.5 mm (WO Cyclone and Spirals) |
| Patherdih | 2.5 MTY | Not yet finalized | | |
| Dugda | 2.5 MTY | | | |
| Bhojudih | 2.0 MTY | | | |

These new washeries have been facilitated with re-washing circuit for the middling produced in their primary circuit; and MoU has already been signed for utilization of rejects produced in these washeries in FBC Power Plants and allied industries (Minutes of IMC Meeting).

With the allocation of new coal blocks for captive usage, new coal washeries in private sector are expected. S&T Mining, a joint venture between Steel Authority of India Limited (SAIL) and Tata Steel Limited has been formed to look into the Mining and Coal Preparation activities of the newly acquired coal blocks to meet the demand of coking coal for their steel plants. Moreover, some American, European, South African and Asian Companies have formed consortiums with Indian Industries to invest in the coal washing sector in India.

With installation of upcoming coal washeries in BCCL, total designed capacity of coking coal washeries in India will be 55.77 MTY by 2020 (Ministry of Coal 2014; Chakraborty et al. 2015; Saxena 2015; Sarkar et al. 2015). The projected production of Coal in BCCL and coal washing targets have been presented in Fig. 2 (Sarkar et al. 2015).

As evident from Table 4 and Fig. 2, present capacity utilization of existing washeries of BCCL is around 20 %. Study is being carried out by the authors for improving techno-economic performance of the washeries. Once the coal washing becomes a profitable proposition, BCCL management will be encouraged to revive the existing washeries for maximum capacity utilization with the objective of washing entire coking coal produced in BCCL. The other companies are expected to follow this and thus ensure utilization of Indigenous Coking Coal with no or substantial reduction in import.

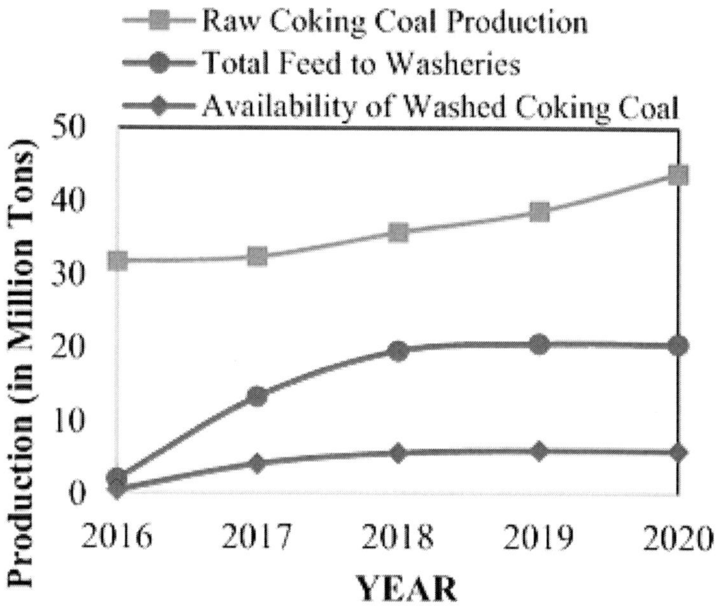


**Figure 2.** Projection of Coking Coal Production in BCCL and washed coal availability to steel sector

## *6* SCOPE FOR RESEARCH AND DEVELOPMENT

As evident from Table 4 and Table 5, with deteriorating quality of coal, the complexity of washing would increase, and operators would resort to 4–5 stages of washing. Outcomes of several experiments carried out on Indian Coal and South African Coal (also of Gondwana Origin) reveal that separate washing of different size fractions have resulted in higher yield (Sapru 2015); and by integrating advanced process control, BATAC Jigs can be utilized to wash difficult-to-wash coal (Mustafi and Nandy 2015). Japan Coal has demonstrated the capability of Variwave Jigs in effectively washing Indian Coal (Koyanagi et al. 2015). FLSmidth has carried out several experiments on Indian Coal and have arrived to a conclusion that washing −13 + 1 mm fraction in Heavy Media Cyclone, −1 + 0.106 mm fraction in spiral concentrator and −0.106 mm fraction in flotation cells will yield better results compared to −13 + 0.5 mm in HM

Cyclone and −0.5 mm by Flotation (Moorhead and Woodruff 2015), which is a regular practice in India. Multotech Process Equipment, SA has conducted several tests on Indian Coal for beneficiation using spiral concentrator; and the results are encouraging (Smith and Erasmus 2015). Dr. Gouricharan, of CIMFR, Dhanbad has patented a novel washing scheme for LVMC Coal by adopting multi-stage, multi-stream coal washing circuit (Gouricharan 2015). However, the complexity of the operation and challenges of maintenance will increase and is likely to pose challenges in smooth operation and maintenance of the plant.

This implies that, there is scope for concerted R&D efforts to find out relatively simpler washing scheme with maximum 2–3 stage of washing, which would not only simplify the operation and maintenance, but also overcome the shortcomings of existing 2–3 stage washing circuits.

There lies scope for comparative study of Indian Coal with Gondwana Coal from other countries in the southern hemisphere. So far, studies have shown that Indian Coal differs significantly from carboniferous coal found in northern hemisphere. However, there is hardly any literature on comparative study of Indian and South African or Australian Coal. This is the reason, coal washing technology used in these countries are accepted as suitable for Indian Coal with little or no modification. Ample opportunity is there for the scientific community to develop indigenous technology designed for difficult washing characteristics of Indian Coal.

Academic & Research institution and technology providers should work in tandem with the industry, as they are ultimately responsible for implementing these technologies. Even after 60 years of Independence, Indian coal washing industry is at its nascent stage. This calls for a Academia-Research Institution and industry consortium to develop new and alternate process routes which are technically easier and superior, environmentally benign and economically sustainable.

# 7 CONCLUSIONS

The existing situation of Indian coal washing Industry is a resultant of gross negligence of the industries to strengthen R&D for development of indigenous technologies. The Industry has relied from the very beginning on imported technology and machinery, ignoring the fact that foreign technologies have been developed taking into account the nature of coal in those countries, which is significantly different from that of Indian Coal, which possess difficult washability characteristics because of its drift origin. Neither the industry nor the government agencies have sponsored an exhaustive R&D project dedicated to coal, despite it being the primary source of energy and infrastructural element in India.

Due to non-availability of such research facilities either at Govt. Research laboratories or Academic Institutions, Indian Scientific community has not yet been able to develop a cause—effect—solution matrix for operational and maintenance practices of Coal Washing Industry. It's one of the principal reasons behind existing condition of public sector Coal Washeries that were installed using imported technology but no indigenous intelligentsia to deal with regular operational and maintenance challenges. Had the academia–industry and research institution been working in tandem, the coal washing scenario in India would have been different.

Thus, attractiveness of Coal Washing Industry in India is high, but also full of challenges. PESTEL analysis of coal washing industry suggests that India is heading towards mandatory coal washing before end utilization. The deteriorating quality of raw coking coal poses challenges before the scientific community and the industry to find out simple and economically viable coal washing scheme for maximum recovery of coking coal for usage in metallurgical sector. There also lies scope for analyzing present condition of existing washeries, especially in public sector, to look for innovative techniques for revival of these washeries.

## REFERENCES


1. Agrawal GC (1976) Public Sector Steel Industry in India, p 76
2. AICC (1965) Indian National Congress. Year of Freedom, vol 19
3. Alam M, Patel JP (2012) Coal washing in BCCL. In: Proceeding of Xstrata—ISM Flotation Workshop, Kolkata, India
4. CCL (Central Coal Fields Limited) History (2015) http://www.ccl.gov.in/comp/our_history.htm. Accessed 20 Oct 2015
5. CEMPDIL report (2016) Coal mining in India: history and perspective.
6. Chakraborty DP, Sen S, Ray PR, Kumar A , Tata Steel Limited (2015) , Augmenting supplies of metallurgical coal to steel mills by upgrading low and medium coking coals, CPSI J, 2015, 66–72
7. CIL (Coal India Limited) History (2015) http://www.coalindia.in/en-us/company/history.aspx. Accessed 20 Oct 2015
8. CMPDI (2015) Completed Projects (120 TPH Modular Washery at Lodna in 1992), Coal beneficiation section in http://scienceandtech.cmpdi.co.in/institution.htm. Accessed 20 Oct 2015
9. CCL (2012) Environmental statement, Kargali Washery, Central Coalfields Limited, 2012–2013
10. Das M (1975) Fantasy of coal nationalization, p 53
11. Ghosh AB (1964) The economics of coal washing in India, pp. 35–36

12. Gouricharan T (2015) Indian coal characteristics—emerging cost effective washing options. CPSI J, pp 125–130

13. Hindustan Steel Limited (1966) Statistics for Iron and Steel Industry in India, p 173

14. Konar BB, Banerjee SB, Choudhury SG, Das NS , Sen K (1998) Coal preparation: CFRI golden jubilee monograph, 1998, pp. 23–24

15. Koyanagi N, Endo H, Kubo Y, JCOAL and NECO (2015) Variwave jig in India, CPSI J, pp 140–144

16. Ministry of Coal (2015) Coal Reserves in India, http://coal.nic.in/content/coal-reserves. 20 Oct 2015

17. Ministry of Coal (2012) Govt. of India, Coal washeries section in http://www.coal.nic.in. Accessed 2012

18. Ministry of Coal (2014) Govt. of India, Minutes of IMC meeting to review the issues related to formulation of policy on coal rejects allocation for washery rejects based power plants held on 08 Oct 2014

19. Ministry of Coal (2014) Govt. of India, Provisional coal statistics 2014–2015

20. Moorhead RG, Woodruff D, Smidth FL (2015) Evaluation of various HMC cleaning strategies to maximize yield of Indian coals, the coal beneficiation 2015 conference, April 2015, pp 79–88

21. Mustafi G, Nandy M (2015) MBE Coal and Mineral Technology India Pvt. Ltd., beneficiation of coal with under-pulsated jigs—is there anything new happening?. CPSI J, 158–164

22. NSO (2015), Ministry of Statistics and Program Implementation, Govt. of India, energy statistics 2015

23. Office of the Chief Economist (2015), Department of Industry and Science, Government of Australia, Coal in India.

24. Rai B, Tiwari R (2009) Coking coals: availability, quality, and beneficiation, proceeding of international conference on coking coals and coke making: challenges and opportunities, January 2009, Ranchi, India

25. Sapru HL (2015) Beneficiation of low volatile medium coking NLW coal in India. CPSI J, April 2015, pp. 55–65, Brief Presentation.

26. Sarkar A, Dutta SK, Patel JP ,Bharat Coking Coal Limited (2015) Augmenting demand of metallurgical coal for Steel Industry: challenges and opportunities, Brief Presentation, the coal beneficiation 2015 conference, April 2015

27. Sarkar GG, Ash PN, Rao SRR (1972) Studies on coal preparation in India, CFRI

28. Saxena BK (2015) Coal India Limited Brief Presentation, CPSI J, Coal Quality related Challenges and CIL's Plan of Action 145–157

29. Sen K (2015) History of coal washing in India. http://www.kalyansen.com/Coal-preparation-in-India-historica-perspective.html. Accessed 20 Oct 2015

30. Singh VB, Guha B (1965) The Coal Mining Industry chapter in economic history of India. Allied Publishers, Chennai, pp 322–323

31. Sinha S (1988) Regional structure, processes, and patterns of development: a case study of Chhota Nagpur Region. Concept Publishing Company, New Delhi, p 74

32. Smith IGT, Erasmus WRJ (2015) Utilization of Spirals when Washing Coal, CPSI J, Multotec Process Equipment, SA, 170–176

33. Tata Steel Limited(2015) Environmental statement for Jamadoba, Bhelatand, West Bokaro - II and West Bokaro - III coal washeries. http://www.tatasteelindia.com/environment-statement.asp. Accessed 20 Oct 2015

# Chapter 6

# Beneficiation of an Indian Non-Coking Coal by Column Flotation

**N. Vasumathi[1] • T. V. Vijaya Kumar[1] • S. Ratchambigai[1] • S. Subba Rao[1] • S. Prabhakar[1] • G. Bhaskar Raju[1]**

*CSIR-National Metallurgical Laboratory Madras Centre, CSIR Madras Complex, Taramani, Chennai 600 113, India*

## ABSTRACT

Beneficiation of non-coking coal is gaining ground in India. It not only reduces the volume of inert content to be transported to the power plant and also lowers the wear in the boiler houses. For special applications such as the fuel for integrated gasification combined cycle plant (IGCC), the ash content in the coal should preferably be below 15 %. Indian coals are characterized by high inter-grown ash content mainly due to 'drift origin' of Gondwana formation in Permian age. This warrants fine grinding of non-coking coal in order to liberate the ash forming minerals from coal macerals. A non-coking coal sample of vitrinite type from India was ground to 44 µm ($d_{80}$) and subjected to column flotation to improve its quality. The non-coking coal analyzing 34.6 % ash, 26.2 % volatile matter, 1.3 % moisture and 37.9 % fixed carbon could be upgraded to a concentrate/froth of

14.83 % ash at 72.18 % yield by optimizing collector and frother dosages and flotation column operating parameters, namely, froth depth, superficial feed velocity and superficial air velocity. The concentrate produced by this process is suitable as fuel for IGCC in coal-to-electricity route.

**Keywords :** Non-coking coal Washability Beneficiation Column flotation Coal gasification

## *1* INTRODUCTION

Coal, accounting for 63 % of the world's total energy content, is the world's most abundant fossil fuel resource. The world supply of crude oil and natural gas is limited and, at projected rate of exploitation is expected to be depleted within the next two or three decades. If worldwide economic development is to continue, alternate fuel resources (such as coal) need to be more fully exploited to meet the anticipated world energy requirement. India is in the threshold of techno-economic washing of coal. So far, large scale beneficiation was resorted to coking coals only for steel making purposes. Now, the stage is set to attract investors to take up beneficiation of non-coking coals also for power plants.

With a growing concern over energy security and sustainability, coupled with concerns about climate change and greenhouse gas emissions from coal combustion, the long term generation of coal based thermal power by India will require the use of cleaner coal and clean coal technologies (CCT). Coal beneficiation is the first and most cost effective step towards satisfying this requirement (Yenigalla 2013). Beneficiation of non-coking coal in India was not given due importance till last 10–12 years due to its low value or not able to meet the cost of the process (Misra et al. 2003). Studies conducted by the International Energy Agency have shown that 1 % increase in ash (generally after passing the 10 % ash level) results in a 1.2 %–1.5 % decrease in boiler availability and a decrease of 0.3 % in boiler efficiency (Konar et al. 1997). To cut down the consumption of oil in existing oil-fired furnaces, coal–oil and coal–water mixture can be used for which low-ash coals recovered from non-coking coals by deep coal cleaning may be of great help. Coal-to-electricity route spans many promising technologies which include integrated gasification combined cycle plant (IGCC). Analysis of data on gasification process has shown that low ash coals perform better for IGCC. The percentage of reserves of inferior non-coking coal is 71 % of total Indian non-coking coal reserves. Therefore, the share of inferior coal production will further increase.

Most Indian coals have a considerable proportion of fixed ash due to 'drift origin' of Gondwana formation in Permian age (Khoury 1981). Most of the thick seams are banded in nature. The bands themselves are usually carbonaceous in nature and very rarely consist of pure sandstones (Mukherjee et al. 1982). These coals are difficult to clean by conventional methods. For applications such as IGCC, these coals need to be ground finer in order to liberate these finely disseminated or inter grown foreign matter. Most of the physical separation techniques have limitations in terms of efficiency of separation at this finer size of grinding. Flotation is the only versatile process in use for fine coal processing in India. Conventional mechanical flotation cells cannot remove all the liberated mineral matter in a single stage. If conventional cells are used, the entrained particles can be removed by refloating the concentrate several times but the cost of multi-stage flotation is often prohibitive. Of late, column flotation has been found to be superior to conventional cells because of a more extended collection zone and improved particle bubble contact efficiency. Keeping this in view, investigations were carried out on a non-coking sample to reduce the ash content from 34.60 % to 15 % in a flotation column.

# 2 MATERIALS AND METHODS

## 2.1 Material characterization

The non-coking coal used in this study was sourced from North Karanpura (Jharkhand, India). The following coal characterization studies were carried out preceding the flotation studies.

## 2.2 Proximate analysis

The proximate analysis of the coal sample is as shown in Table 1. It could be observed that the high ash content in this coal warrants processing for its utilization.

**Table 1.** Proximate analysis (air dry basis) of non coking-coal sample (%)

| $A_{ad}$ | $M_{ad}$ | $V_{ad}$ | $FC_{ad}$ |
|---|---|---|---|
| 34.6 | 1.3 | 26.2 | 37.9 |

## 2.3 Petrographic analysis

The petrographic analysis of the coal sample was carried out and the results are as shown in Table 2. Its mean maximum reflectance was found to be 0.71 %.

**Table 2.** Petrographic analysis of non-coking coal sample (%)

| Vitrinite | Semi-vitrinite | Liptinite | Inertinite | Mineral matter |
|---|---|---|---|---|
| 38.7 (46.5) | 0.1 (0.2) | 11.9 (14.4) | 32.4 (38.9) | 16.9 |

The values in parenthesis are on 'mineral-free' basis. Earlier researchers established the following floatability order of the petrographic components (Brady and Gauger 1940; Arnold and Aplan 1989; Gosiewska et al. 2002; Ding 2009).

Vitrinite > Exinite > Inertinite > Inorganic mineral matter

This floatability order indicates the theoretical possibility of the separation of different petrographic components of the present coal sample.

## 2.4 Size analysis

The size analysis of coal was carried out and the results are as shown in Table 3.

**Table 3.** Size analysis of coal ($d_{80}$: 2108 μm)

| Size (mm) | Weight retained (%) | $A_{ad}$ (%) | Ash distribution (%) |
|---|---|---|---|
| +6 | 1.00 | 39.12 | 1.14 |
| −6 + 2.8 | 27.52 | 24.93 | 20.00 |
| −2.8 + 1.68 | 22.82 | 31.25 | 20.79 |
| −1.68 + 0.85 | 25.17 | 38.16 | 28.00 |
| −0.85 + 0.5 | 14.43 | 46.25 | 19.46 |
| −0.5 + 0.212 | 6.71 | 39.18 | 7.66 |
| −0.212 | 2.35 | 43.08 | 2.95 |

From the above table, it is clear that ash is distributed predominantly in the size range of −6 + 0.5 mm. The percent ash distribution value in the individual size fractions in this size range is around 20 %. The coal was further subjected to crushing in rolls for liberation of ash forming minerals.

## 2.5 Washability study

The coal was sieved at 0.5 mm and the −0.5 mm fraction of coal was segregated. The +0.5 mm coal was subjected to heavy media separation (HMS) in successive baths of increasing specific gravity. The medium of desired density is obtained by using organic liquid mixtures of bromoform, tetrachloroethylene and benzene. From the HMS study, the washability curves are plotted as shown in Fig. 1. From these, it is evident that 70 % of the coal size fraction (-6 + 0.5 mm) could be obtained as float with 15 % ash content by subjecting it for heavy medium separation at 1.70 specific gravity. The rejects amount to 30 wassaying 70 % ash (as shown in black dotted lines).

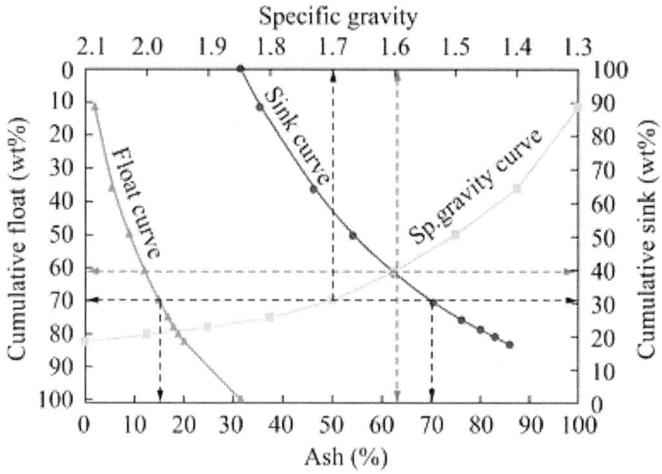


**Figure 1.** Washability curves for -6 + 0.5 mm fraction of coal

## 2.6 Flotation column

The automated laboratory flotation column with an internal diameter of 75 mm, designed and developed by CSIR-NML Madras Centre was used in the present study to know the effect of its operating parameters on the process. The schematic details of laboratory column were shown in Fig. 2. Electronically controlled metering pumps were used to feed and discharge the slurry. Slurry/froth interface was maintained using differential pressure transmitter (DPT). The output signal of the DP cell was looped with the stroke controller of the tailings pump so that the pumping rate could be automatically varied to maintain the interface level at a fixed froth depth. Sintered ceramic tube air sparger is used with 30 % porosity. Purge rotameter with a differential pressure regulator was used to control the flow of air to the column from a compressor. The column was filled with water and stabilized at required froth depth. The coal slurry at ~10 % solids by weight was conditioned initially with diesel (collector) in a conditioner for 3 min. In the second stage, MIBC (frother) was added and conditioned for another 3 min. This slurry was pumped to the column. The column was allowed to run for a minimum period of 3–4 residence times. Samples were drawn under near-steady state conditions. Both the process parameters and column operating conditions were recorded before collecting the samples. Samples were analyzed for ash content by standard method.

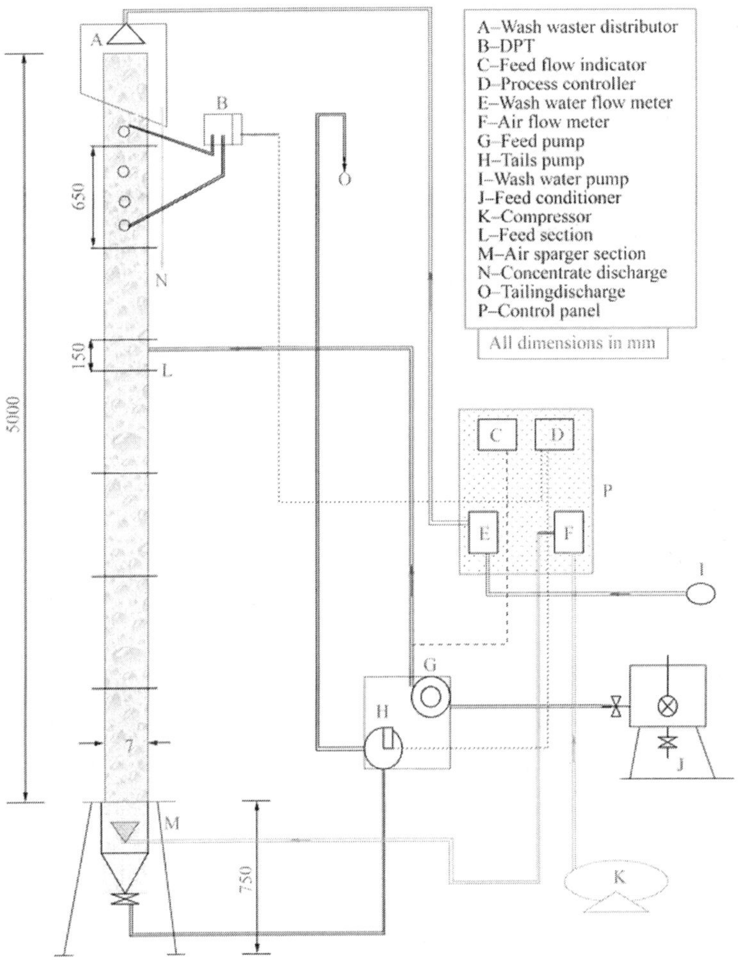


**Figure 2.** Schematic diagram of CSIR-NML laboratory flotation column

# 3 RESULTS AND DISCUSSION

Mechanized coal mining techniques generate large amount of fine particles that have to be recovered in coal washing plants. Fine coal processing has always been a problematic and costlier than the cleaning of coarse coal (Andrew Mohns 1997). Froth flotation technique is an effective and best available technique for cleaning fine coal with particle size less than approximately 0.5 mm (Bensley and Nicol 1985; Yoon et al. 2002; Koca et al. 2010). Froth flotation, which exploits the difference in surface properties of minerals, is recognized as an efficient method for coal cleaning and processing of minerals. The efficiency in separating the impurities from coal by flotation is basically determined by the relative hydrophobicity of coal and the associated gangue (Laskowski 2001; S Dey et al. 2013). In froth flotation, coal particles are often subjected to a suitable hydrocarbon oil treatment (Boylu and Laskowski 2007) to alter their hydrophobicity, enhance recovery, and/or improve selectivity (Peng 1996). Conventionally, the collector used in most of the Indian coal washeries is diesel oil in combination with different frothers. The dosage of collector and frother has significant effect on flotation performance (Dey and Pani 2012). Therefore, optimization of grind, collector and frother dosages was carried out followed by that of the operating parameters of flotation column.

## 3.1 Grind optimization studies

To study the maximum achievable separation of combustible material, the coal sample was subjected to stage crushing to obtain −1.68 mm product which was then fed to the ball mill. A representative sample was drawn for particle size and ash analysis and results are tabulated as Table 4.

**Table 4.** Size analysis of feed to ball mill ($d_{80}$: 1102 µm)

| Size (µm) | Weight retained (%) | $A_{ad}$ (%) | Ash distribution (%) |
|---|---|---|---|
| + 850 | 44.15 | 33.49 | 43.67 |
| −850 + 425 | 33.87 | 35.40 | 35.41 |
| −425 + 212 | 10.24 | 31.12 | 9.41 |
| −212 + 106 | 6.35 | 30.43 | 5.71 |
| −106 | 5.39 | 36.42 | 5.80 |

The above sample was subjected for wet grinding in a laboratory ball mill and the grind variation studies were carried out for various grinding times (15, 20, 25, 30 and 35 min) at the pulp density of 60 % solids by weight. Batch flotation tests were carried out on ground coal samples to optimize the 'mesh of grind' to recover the maximum combustibles. Bench scale flotation tests were performed in a D12 Denver Flotation Machine. The ground samples were subjected to flotation using diesel as collector and MIBC as frother. The combustible recovery (CR), ash rejection (AR) and efficiency index (EI) were calculated by using the following standard formulae (Vasumathi et al. 2013).

$$\%CR = \left[\frac{M_c(100 - A_c)}{M_f(100 - A_f)}\right] \times 100$$

$$\%AR = \left[1 - \left(\frac{M_c A_c}{M_f A_f}\right)\right] \times 100$$

$$EI = (CR + AR) - 100$$

where $A_c$ is the ash in concentrate, %; $A_f$ is the ash in feed, %; $M_c$ is the mass of concentrate and $M_f$ is the mass of feed.

The results in Table 5 reveal that it is possible to obtain a froth/concentrate of 56 % yield with less than 12 % ash by flotation at

$d_{80}$ of 44 µm corresponding to 30 min. gGrind. Also, greater yield of 77 % of froth with 19.7 % ash content can be achieved without cleaning the rougher concentrate. Hence, it could be possible to obtain required 15 % ash with maximum yield at this grind by optimizing the reagent dosages.

**Table 5.** Flotation of coal at different grinding times (roughing followed by cleaning)

| Grinding time (min) | Product | Yield (%) | $A_{ad}$ (%) | CR (%) | AR (%) | EI |
|---|---|---|---|---|---|---|
| 15 | F C | 8.31 | 5.91 | 11.65 | 98.51 | 10.16 |
| | Cl.2-T | 33.85 | 11.50 | | | |
| | Cl.1-T | 34.46 | 34.61 | | | |
| | P T | 23.38 | 71.05 | | | |
| 20 | F C | 51.51 | 13.88 | 68.95 | 79.96 | 48.91 |
| | Cl.T | 28.43 | 45.11 | | | |
| | P T | 20.07 | 78.22 | | | |
| 25 | F C | 42.67 | 12.56 | 57.19 | 84.59 | 41.78 |
| | Cl T | 31.67 | 37.27 | | | |
| | P T | 25.67 | 68.59 | | | |
| 30 | F C | 56.10 | 11.77 | 72.99 | 79.49 | 52.48 |
| | Cl T | 20.91 | 41.08 | | | |
| | P T | 23.00 | 73.90 | | | |
| 35 | F C | 51.86 | 12.31 | 68.40 | 80.95 | 49.35 |
| | Cl T | 26.10 | 42.09 | | | |
| | P T | 22.04 | 73.24 | | | |

*Note FC* final concentrate, *Cl T* cleaner tailings, *PT* primary tailings.diesel is 2.49 kg/t, MIBC is 0.66 kg/t, head ash of coal is 34.60 %

## 3.2 Column flotation studies

The collector and frother dosages and column operating parameters were optimized on laboratory scale flotation column. The grind of the coal was maintained the same ($d_{80}$: 44 µm) in all the following tests.

### 3.2.1 Optimization of collector (diesel) and frother (MIBC) dosages

The test conditions and the results on the effect of variation of diesel and MIBC dosages are given in the Tables 6 and 7, respectively.

**Table 6.** Effect of diesel on column flotation of coal

| Diesel (kg/t) | Product | Yield (%) | Ash (%) | Ash distribution (%) | EI |
|---|---|---|---|---|---|
| 2.55 | Concentrate | 73.94 | 16.78 | 37.31 | 54.9 |
| | Tailings | 26.06 | 80.00 | 62.69 | |
| 2.87 | Concentrate | 78.10 | 17.34 | 43.25 | 53.1 |
| | Tailings | 21.90 | 81.14 | 56.75 | |
| 3.19 | Concentrate | 77.66 | 21.28 | 50.97 | 39.52 |
| | Tailings | 22.34 | 71.17 | 49.03 | |
| 3.51 | Concentrate | 88.78 | 21.35 | 67.13 | 30.18 |
| | Tailings | 11.22 | 82.72 | 32.87 | |

*Note* MIBC dosage is 0.65 kg/t, froth depth is 600 mm, superficial air velocity is 0.78 cm/s, superficial feed velocity is 0.57 cm/s, head ash of coal is 34.60 %

**Table 7.** Effect of MIBC on column flotation of coal

| MIBC (kg/t) | Product | Yield (%) | $A_{ad}$ (%) | Ash distribution (%) | EI |
|---|---|---|---|---|---|
| 0.33 | Concentrate | 53.46 | 13.26 | 23.24 | 43.48 |
| | Tailings | 46.54 | 50.30 | 76.76 | |
| 0.49 | Concentrate | 78.26 | 21.31 | 47.91 | 46.56 |
| | Tailings | 21.74 | 83.40 | 52.09 | |
| 0.57 | Concentrate | 76.11 | 21.82 | 46.60 | 45.86 |
| | Tailings | 23.89 | 79.65 | 53.40 | |
| 0.65 | Concentrate | 88.78 | 21.35 | 67.13 | 30.18 |
| | Tailings | 11.22 | 82.72 | 32.87 | |

*Note* Diesel dosage is 2.87 kg/t, froth depth is 600 mm, superficial air velocity is 0.78 cm/s, superficial feed velocity is 0.57 cm/s , head ash of coal is 34.60 %

It could be observed that as the diesel dosage is increased from 2.55 kg/t to 3.51 kg/t, the yield of the froth increased from 73.94 % to 88.78 % with ash in it increasing from 16.78 % to 21.35 %. At the same time, efficiency index of separation decreases from 54.90 to 30.18. This could be attributed to the non-selectivity of diesel at higher dosages and physical entrainment/entrapment of ash forming minerals in the froth resulting in higher yield and ash content in the froth.

The dosage of frother plays vital role as it affects the bubble size distribution and mean bubble size. In flotation, the bubble is the driving force and it determines the separation efficiency. The flotation results at various MIBC dosages and at a constant diesel dosage are presented in Table 7. The results clearly indicate that MIBC plays very important role in coal flotation. At constant diesel dosage, flotation concentrate yield increased as the MIBC dosage increased.

## 3.3 Optimization of column operating parameters

The following column parameters namely air flow rate, feed flow rate, froth depth, wash water and % solids of feed were optimized for non-coking coal flotation.

### 3.3.1 Effect of air flow rate

Generally, superficial air velocity should be as large as possible to ensure a high throughput. At the same time, if the superficial air velocity is too high, the flow pattern will be disturbed and there is every possibility to lose bubbly swarm. It is to be noted that column should operate in a bubbly flow regime, where gas hold-up varies linearly with air rate. A proper combination of gas rate and bubble size will generally provide a gas holdup in the flotation pulp in the range of 15 %–18 %.

The effect of superficial air rate on the flotation of coal is presented in Table 8. It is observed that as the superficial air rate increases there is no increase in the clean coal yield. But at higher superficial air rates the ash content in clean coal decreased marginally. This may be due to the reduction in gangue entrainment. The effect of superficial air rate on entrainment in froth may be explained as follows: the ash content in the clean coal is reduced at higher superficial air velocity because of increase in the bubble size in the froth zone. In the present investigation it was observed that the superficial air velocity of 0.78 cm/s was found to be optimum to obtain better grade and recovery.

**Table 8.** Effect of air rate on column flotation of coal

| Superficial air velocity (cm/s) | Product | Yield (%) | Ash (%) | Ash distribution (%) | EI |
|---|---|---|---|---|---|
| 0.39 | Concentrate | 78.63 | 20.14 | 46.88 | 47.95 |
| | Tailings | 21.37 | 83.96 | 53.12 | |
| 0.59 | Concentrate | 77.95 | 21.72 | 46.66 | 49.12 |
| | Tailings | 22.05 | 87.79 | 53.34 | |
| 0.78 | Concentrate | 78.10 | 17.34 | 43.25 | 53.10 |
| | Tailings | 21.90 | 81.14 | 56.75 | |
| 0.98 | Concentrate | 73.14 | 18.74 | 39.31 | 51.92 |
| | Tailings | 26.86 | 78.77 | 60.69 | |

*Note* Diesel dosage is 2.87 kg/t, MIBC dosage is 0.65 kg/t, froth depth is 600 mm, superficial feed velocity is 0.57 cm/s, head ash of coal is 34.60 %

### 3.3.2 Effect of feed flow rate

Change in feed flow rate usually affects grade and recovery. In the case of coal flotation, the increase in volumetric feed flow rate results in decrease of residence time of slurry which will affect the recovery. Enough residence time is to be maintained so that entire coal will be collected and separated. Feed flow velocities were varied from 0.33 to 0.87 cm/s by manipulating the speed of the variable frequency driven feed pump and the results were presented in Table 9.

**Table 9.** Effect of Feed velocity on column flotation of coal

| Superficial feed velocity (cm/s) | Product | Yield (%) | $A_{ad}$ (%) | Ash distribution (%) | EI |
|---|---|---|---|---|---|
| 0.33 | Concentrate | 80.39 | 17.78 | 48.57 | 45.10 |
| | Tailings | 19.61 | 77.18 | 51.43 | |
| 0.41 | Concentrate | 79.58 | 16.99 | 45.56 | 48.39 |
| | Tailings | 20.42 | 79.11 | 54.44 | |
| 0.49 | Concentrate | 76.67 | 18.41 | 43.23 | 49.65 |
| | Tailings | 23.33 | 79.46 | 56.77 | |
| 0.57 | Concentrate | 78.10 | 17.34 | 43.25 | 53.10 |
| | Tailings | 21.90 | 81.14 | 56.75 | |
| 0.65 | Concentrate | 79.26 | 21.19 | 49.98 | 44.09 |
| | Tailings | 20.74 | 81.04 | 50.02 | |
| 0.87 | Concentrate | 78.12 | 20.28 | 45.49 | 50.04 |
| | Tailings | 21.88 | 86.75 | 54.51 | |

*Note* Diesel dosage is 2.87 kg/t, MIBC dosage is 0.65 kg/t, froth depth is 600 mm, superficial air velocity is 0.78 cm/s, head ash of coal is 34.60 %

As the superficial feed velocity increases, generally the weight recovery would decrease owing to the decrease in pulp residence time. On the other hand the quality of the concentrate generally increases due to less residence time of froth in the froth zone. From the results it is observed that the yield of clean coal is not much affected by increasing the superficial feed velocity which may be due to the natural floatability of coal. But at higher feed velocities, the ash content is increased in the clean coal because of less residence time of froth in the cleaning zone. Generally, good coal doesn't require larger residence times, but in the present case, the quality of clean coal is better at higher residence time i.e., at feed velocity of 0.41 cm/s. The calculated residence time at this feed velocity is about 18–20 min.

### 3.3.3 Effect of froth depth

Floatability characteristics are also another factor which decides the height of the flotation zone. If the material is having higher floatability like graphite and coal, the froth height increases and does not affect much on yield but simultaneously the concentrate grade increases as demonstrated by the present studies. Further, it has been seen that by increasing the froth depth, the entrained gangue minerals in the concentrate reduce significantly due to coalescence of bubbles, positive bias water and reduction of feed water in the concentrate. Experiments were conducted at different froth depths ranging from 500 mm to 900 mm and the results obtained were presented in Table 10. The rejection of entrained particles, in this case, inorganic ash forming minerals, depends on froth depth. Due to high superficial air and feed velocities, fine and ultra fine sized ash forming particles might be entrained into the froth phase. If there is no enough froth depth, these particles will be carried along with the clean coal and thus the quality may be affected and the froth depth of 600 mm is optimized for further experiments.

**Table 10.** Effect of froth depth on column flotation of coal

| Froth depth (mm) | Product | Yield (%) | $A_{ad}$ (%) (%) | Ash distribution (%) | EI |
|---|---|---|---|---|---|
| 500 | Concentrate | 78.53 | 21.13 | 48.70 | 45.24 |
| | Tailings | 21.47 | 81.41 | 51.30 | |
| 600 | Concentrate | 78.10 | 17.34 | 43.25 | 53.10 |
| | Tailings | 21.90 | 81.14 | 56.75 | |
| 700 | Concentrate | 76.02 | 21.10 | 47.26 | 43.53 |
| | Tailings | 23.98 | 74.65 | 52.74 | |
| 800 | Concentrate | 79.35 | 16.41 | 44.54 | 49.21 |
| | Tailings | 20.65 | 78.52 | 55.46 | |
| 900 | Concentrate | 70.78 | 14.50 | 31.48 | 58.30 |
| | Tailings | 29.22 | 76.45 | 68.52 | |

*Note* Diesel dosage is 2.87 kg/t, MIBC dosage is 0.65 kg/t, superficial feed velocity is 0.57 cm/s, superficial air velocity is 0.78 cm/s, head ash of coal is 34.60 %

It could be seen from the above results that the clean coal yield reduced at higher froth depths. This could be due to froth drop back into slurry phase across the froth/slurry interface at relatively higher froth depth. At the same time the ash content in the clean coal is reduced gradually as the froth depth increases.

### 3.3.4 Effect of feed % solids (pulp density)

In order to examine the effect of pulp density on the flotation of coal, tests were conducted at various pulp densities and the results of the same are given in Table 11.

**Table 11.** Effect of pulp density on column flotation of coal

| % Solids by weight | Product | Yield (%) | $A_{ad}$ (%) | Ash distribution (%) | EI |
|---|---|---|---|---|---|
| 4.76 | Concentrate | 72.18 | 14.83 | 35.44 | 52.63 |
| | Tailings | 27.82 | 70.09 | 64.56 | |
| 9.09 | Concentrate | 78.10 | 17.34 | 43.25 | 53.10 |
| | Tailings | 21.90 | 81.14 | 56.75 | |
| 13.04 | Concentrate | 87.94 | 20.14 | 66.53 | 29.18 |
| | Tailings | 12.06 | 73.88 | 33.47 | |
| 16.67 | Concentrate | 63.17 | 19.50 | 33.36 | 47.25 |
| | Tailings | 36.83 | 66.80 | 66.64 | |

*Note* Diesel dosage is 2.87 kg/t, MIBC dosage is 0.65 kg/t, froth depth is 600 mm, superficial feed velocity is 0.57 cm/s, superficial air velocity 0.78 cm/s, head ash of coal is 34.60 %

The decrease in selectivity in flotation of clean coal was observed when the pulp density is increased from 4.76 % to 16.67 %. But the efficiency of separation is better between 5 % and 9 % solids.

## 4 CONCLUSIONS

A non-coking coal of vitrinite variety and analyzing 34.6 % ash, 26.2 % volatile matter, 1.3 % moisture and 37.9 % fixed carbon was subjected for flotation and column flotation studies. Concentrate/froth of 14.83 % ash at 72.18 % yield could be achieved at optimum process parameters such as 'mesh of grind' ($d_{80}$: 44 µm), collector dosage (diesel: 2.87 kg/t) and frother dosage (MIBC: 0.65 kg/t) and flotation column operating parameters like froth depth (600 mm), superficial feed velocity (0.57 cm/s), superficial air velocity (0.78 cm/s) and feed pulp consistency of 9 % solids by weight. Out of these parameters, froth depth and slurry feed percent solids have significant effect on the process. The concentrate produced by this process is suitable as fuel for IGCC in coal-to-electricity route.

## ACKNOWLEDGMENTS

The authors are thankful to The Director, CSIR-NML for his encouragement and support for publishing this work.

## REFERENCES

1. Andrew Mohns C (1997) Effect of particle size on coal flotation kinetics (PhD thesis). Queen's University, Kingston:pp1–4

2. Arnold BJ, Aplan FF (1989) The hydrophobicity of coal macerals. Fuel 68(5):651–658

3. Bensley CN, Nicol SK (1985) The effect of mechanical variables on the flotation of coarse coal. Coal preparation 1:189

4. Boylu F, Laskowski JS (2007) Rate of water transfer to flotation froth in the flotation of low-rank coal that also requires the use of oily collector. Int J Miner Process 83:125–131

5. Brady GA, Gauger AW (1940) Properties of coal surfaces. J ind eng chem 32:1599–1604

6. Dey S, Pani S (2012) Effective processing of low-volatile medium coking coal fines of Indian origin using different process variables of flotation. Int J Coal Prep Util 32:253

7. Ding LP (2009) Investigation of bituminous coal hydrophobicity and its influence on flotation. Energy fuels 23:5536–5543

8. Gosiewska A, Drelich J, Laskowski JS, Pawlik M (2002) Mineral matter distribution on coal surface and its effect on coal wettability. J Colloid Interface Sci 247(1):107–116

9. Khoury DL (1981) Coal Cleaning Technology. NDC, Park Ridge

10. Koca S, Bektas Y, Koca H (2010) Contact angle measurement on lignite surface. Proceedings of the XXV International Mineral Processing Congress (IMPC) Brisbane 2049

11. Konar BB, Banerjee SB, Chaudhuri SG, Choudhury A, Das NS, Sen Kalyan (1997) In: Coal Preparation. Allied Publishers Ltd New Delhi, CFRI Golden jubilee monograph

12. Laskowski JS (2001) Coal flotation and fine coal utilization. Elsevier Science, Amsterdam, p 225

13. Mukherjee AK, Chatterjee CN, Ghose S (1982) Coal Resources of India:it's formation, distribution and utilisation. Fuel Sci Technol 1:19–34

14. Peng FF (1996) Surface energy and induction time of fine coals treated with various levels of dispersed collector and their correlation to flotation responses. Energy fuels 10:1202–1207

15. Shobhana Dey, Gyana Manjari Paul, Santosh Pani (2013) Effective flotation of weathered coal using frother blend. In: Coal preparation technology CSIR - CIMFR Jharkhand

16. Vasumathi N, Vijaya Kumar TV, Subba Rao S, Prabhakar S, Bhaskar Raju G, Shiva Kumar S, Raman Uma (2013) Eco-friendly and cost effective reagent Sokem 590C for coal flotation. Int J Eng Res 2(7):418–433

17. Yenigalla GR (2013) Coal Beneficiation: A road to clean coal Technology. In CSIR - CIMFR Jharkhand India: Coal Preparation Technology Proceedings

18. Yoon RH, Luttrell GH, Asmatulu R (2002) Extending the upper particle size limit for coal flotation. J South African Inst Mining Metallurgy 102(7):411–415

# Chapter 7

## The Technical Analysis on Upgrade and Coupling of Low-Quality Coal

**Bin Zhao[1,2], Qiang Yao[1], Shan Lu[1], Qinggong Wang[2], Junfu Lv[2]**

[1]*Key Laboratory for Modem Metallurgy Technology of Ministry of Education, Hebei United University, Tangshan, China*

[2]*Key Laboratory for Thermal Science and Power Engineering of Ministry of Education, Department of Thermal Engineering, Tsinghua University, Beijing, China*

## ABSTRACT

Experimental study was conducted with regard to upgrading low-quality coal by low temperature pyrolysis technology and fluidization classification technology, which laid the theoretical foundation for research and development of low-quality coal upgrading process. Firstly, a pyrolysis pilot experiment of long-flame coal was performed and the optimum pyrolysis conditions were found. When the final pyrolysis temperature was 450°C and the residence time was 10 minutes, the content of semicoke volatile dry ash-free was about 22.89% and the content of semicoke ash was about 16.1%. More than 5% of the ash needs to be removed before entering the pyrolysis fluidized bed. Thus, classification tests were carried out to pick out the coal particles with the target size of 500 μm, using a gas-solid fluidized bed coal picker. It was found that the separating effect of coal particles was the most satisfactory when the inlet velocity at the bottom was 3.27 m/s. The percentage of particles with diameters less than 500 μm was as low as 28.7% in the coarse samples. Based on the test results, a novel process of low-quality coal upgrade and coupling was proposed, which realized the sorting, grading, drying and pyrolysis of low-quality coal through the multi-stage fluidized bed integrated process.

**Keywords:** Low-Quality Coal, Low Temperature Pyrolysis, Fluidized Bed Particle Classification

# 1. INTRODUCTION

Coal is a main product of China's energy consumption. Under the current energy situation, to develop coal clean technology and to improve the utilization of coal are of great significance to alleviate the pressure of the future energy. In 2012, China's total coal reserves reached 2298.9 tons. In the recoverable coal reserves, low rank coal accounted for about 40%. A plenty of coal in low rank coal has high impurity (high ash, high water and high sulfur) characteristics, which is called the low quality coal.

Li Zhen, Yu Wei, Yang Chao and Zhou Anning pointed out that the treatment of low rank coal slime is an effective way on basis analysis in the comprehensive utilization of low rank coal and coal slime [1] . Gu Xiaoyu had an experimental study on two kinds of Shenhua low rank coal of thermal modification qualitative processing researches. The results showed that the thermal modification could significantly improve the coal quality characteristic of low rank coal, eventually getting low moisture and high calorific value and improving the ignition quality of modified products [2] . The Chinese Academy of Sciences Institute of the Development Process of Coal Topping Process (BT process) which uses fast pyrolysis, rapid separation and rapid cooling realizes priority use of tar oil and gas and solid products such as carbocoal which is used to generate electricity [3] . In the 1970s, the rapid development of fluidized bed pyrolysis process for a variety of bituminous coal and lignite pyrolysis were researched by Australia's CSIRO. The reactor bed of circulating fluidized bed technology is composed by the grit with 0.3 - 1 mm. The residence time is short, and the energy from the flue gases of pyrolysis by the electric heater and the liquefied petroleum gas and nitrogen gas preheated air fuel formation [4] .

This paper is based on low temperature pyrolysis and fluidized bed coal particles classification tests of fluidized bed. Based on the results, the process integrated design of multi-stage fluidized bed was provided for the sorting, grading, drying and pyrolysis of low-quality coal in the integrated iron system. On one hand, this design is conducive to complementary chiasma using of the waste heat in working procedures of iron making. On the other hand, the process integration design is in favour of the comprehensive utilization of low-quality coal. Furthermore, this new technology will reduce the cost of irons effectively, thus demand pressure of higher-order coal in steel industry will slow down.

## 2. CHARACTERISTICS OF LOW QUALITY COAL

The main features of low-quality coal are high impurity. High impurity content will not only reduce the quality of coal, but also increase transportation costs and processing costs. If low-quality coal is directly used, it is easy to cause environmental pollution and energy waste. Taking sintering and coking as an example, with excessive moisture content, particularly with the high internal waters content in the coal, the water will break away from the surface of coal particles in the initial stage of accelerated warming, which is easy to cause the pores in the surface of coal particles ditch and even collapse and the sinter or coke exhibit more cracks. Then the mechanical strength of the product is reduced, and energy is more consumed during the heating process. Similarly, when the coal particles conduct to the liquid phase reaction by heating it, the higher ash content will impact the liquid flow resistance and reduce the bonding effect. While the mechanical strength of sinter and coke depends mainly on the full extent of liquid-phase reaction, the higher ash content will exacerbate pollution emissions. Since sinter and coke are both to enter blast furnace for iron making, the higher sulfur content will not only increase the sulfide emissions but also lead to the phenomenon of hot crisp iron. When dissolving in liquid iron in the form of ferrous sulfide, the sulfur will be enriched in the grain boundary in the solidification process, which will form low melting point eutectic, weaken the binding force between the grains and cause brittle cracking phenomenon.. For the elimination of low-quality coal high impurity process, preliminary mention quality treatment often cannot achieve the purpose of improving quality, but also requires combing the deep technology of quality, adjusting process reasonable, and using multi-stage processing means to achieve the ultimate upgrading.

In summary, to achieve efficient use of low-quality coal, it is necessary to carry out upgrading process from the source, that is, starting from the sorting and grading, not only to the reasonable control of component content in coal, but also on the particle size classification of coal particles to adapt to the needs of deep quality on low-quality coal.

# 3. KEY TECHNOLOGY OF LOW-QUALITY COAL UPGRADING

## 3.1. Low Temperature Pyrolysis

Using long-flame coal as pyrolysis experiment of coal, coal quality analysis results are shown in Table 1. As can be seen from Table 1, the

**Table 1.** Analysis of test coal.

| Analysis [%] | Value |
|---|---|
| Mt | 19.5 |
| Mad | 11.8 |
| Ad | 17.07 |
| Vdaf | 37.53 |
| FCdaf | 62.47 |
| Qnet,ar ($MJ\cdot kg^{-1}$) | 18.83 |
| Cad | 56.53 |
| Had | 4.9 |
| Oad | 10.98 |
| Nad | 0.45 |
| Sad | 0.28 |

Mt is for total moisture; Mad is for inherent moisture; Ad is for dry base ash; Vdaf is for volatile content on dry ash free basis; FCdaf is for fixed carbon content on dry ash free basis; Qnet,ar is the net calorific power of as received basis; Cad is for air-dried basis carbon; Had is for air-dried basis hydrogen; Oad is for air-dried basis oxygen; Nad is for air-dried basis nitrogen; Sad is for air-dried basis sulfur.

internal waters, volatile component and ash content are high. If we direct sintering or coking, it will reduce the coke strength. Meanwhile, because of higher volatile content of coal reduces the safety of production process, it also needs pre-removal. In summary, the coal used in the pyrolysis test program should be pre-dried to remove the remaining portion of the water to 5%.

The muffle furnace was used within the pyrolysis test, taking 1 g ± 0.0001 g of coal (diameter < 0.2 mm) placed in a crucible, then set the final pyrolysis temperature at 450°C, 500°C, 550°C and 600°C, respectively, residence time for 5 min, 10 min, 30 min, 50 min, 70 min and 90 min. Various conditions of semi-coke volatile content are shown in Figure 1, Figure 2 is ash content of semi-coke.

From the analysis of Figure 1, with the improvement of terminal pyrolysis temperature, the energy obtained by the broken bonds the functional group increased and the rate of initial stage of devolatilization is increasing, the precipitation amount of volatiles is proportional to the final pyrolysis temperature. When the final pyrolysis temperature of 450°C, the residence time is 10 min, the dry ash-free basis of semicoke volatile content is about 22.89%, which can meet the injection coal blending and general transport requirements of metallurgical coke. Figure 2 shows that, with the final pyrolysis temperature increases, the corresponding increase in the ash content. In the general requirements, the coal ash content of metallurgical coking is less than 12%, sintering coal generally require less than 15%. When the final pyrolysis temperature is 450°C, the residence time of 10 min, the ash content of semicoke is about 16.1%. Therefore, in the upgrading process design, it needs ash handling for coal before entering into pyrolysis fluidized bed. In summary, the optimum condition of pyrolysis fluidized bed upgrading: final pyrolysis temperature of 450°C, the residence time of 10 min, remove more than 5% of ash before entering the pyrolysis fluidized bed.

### 3.2. Fluidized Classification

Gas-solid fluidized bed coal washer as the main body in the test equipment to conduct coal particle classification experiment research in a certain steel company. Tests mainly conducted on the effect of fluidization velocity on classification efficiency. Through the analysis of different conditions, the content of coal particle with its size less than 500 μm, in the lower refined coal to further determine the best classification conditions. The particle size distribution is shown in Table 2. The concentration distribution of particle size less than 500 μm is shown in Figure 3.

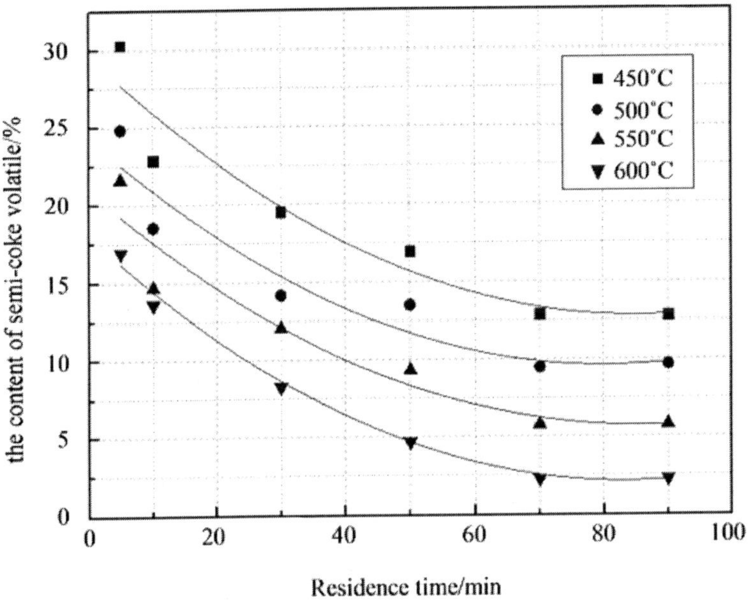


**Figure 1.** The content of semi-coke volatile under various conditions.

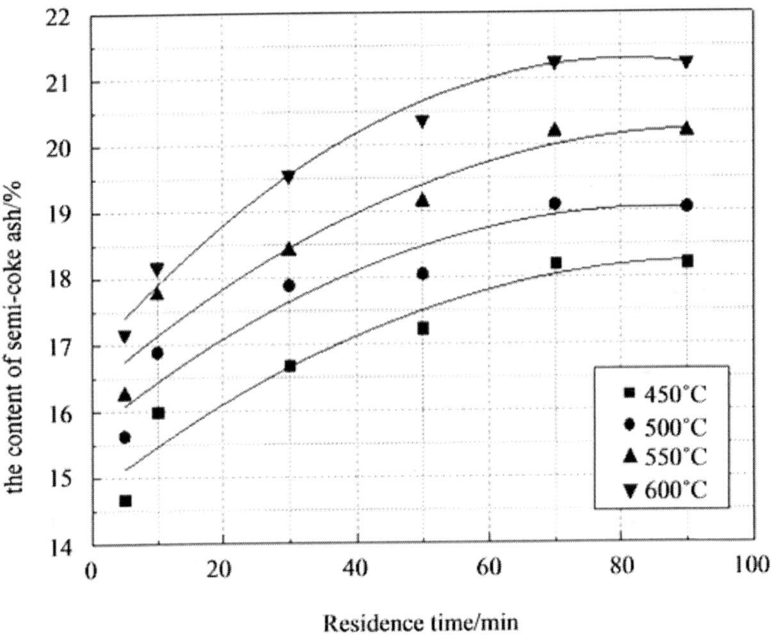


**Figure 2.** The content of semi-coke ash under various conditions.

**Table 2.** The initial parameter of coal particle size distribution in the experiments.

| Particle size [μm] | Accumulative contents [%] |
|---|---|
| 100 | 7.57 |
| 200 | 17.98 |
| 500 | 60.41 |
| 1000 | 72.61 |
| 2000 | 90.43 |
| 3000 | 100 |

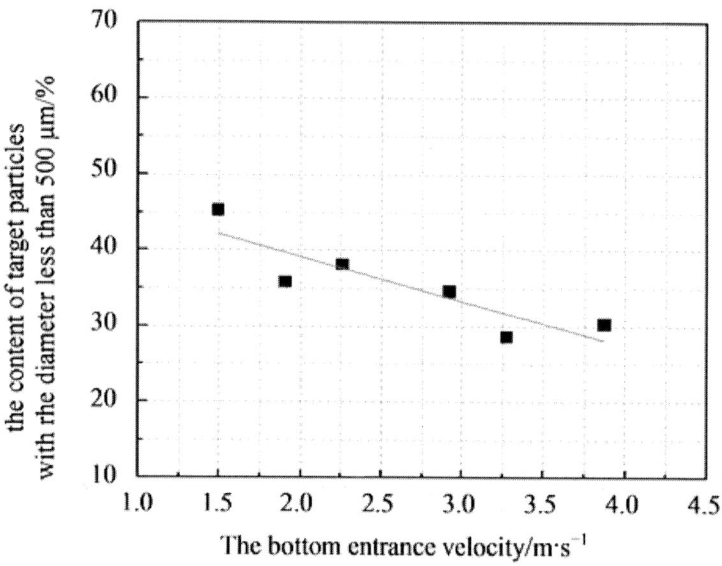


**Figure 3.** The content of particles with the diameter less than 500 mm under different bottom entrance velocity.

It can be seen from Figure 3, with the increase of fluidization velocity, the content of coal particle with its size less than 500 μm presents a descending tendency in lower refined coal. Classification has the best effect when the bottom entrance velocity is 3.27 m/s, the content of coarse coal particles the size less than 500 μm reduced to 28.7%.

# 4. LOW QUALITY COAL AND IRON THE QUALITY PROCESS

## 4.1. Identified Ideas of Upgrading in Low-Quality Coal

Accompanied by iron-making process, there have been migration and transformation of energy. And in the process of practical production, waste heat can be serious wasted, but the cause of this phenomenon is not just the quality of equipment and operational level. In order to further reduce the consumption, we should choose two points to analyze: one is the initial resource rationalization. It strictly controls the quality of materials which enter into the process system, especially the quality of coal, because coal, grain size and other factors will directly restrict the efficiencies of heat utilities. The other is rational process. On the basis of meeting the needs of normal production, the process should also take the heat recovery efficiency and continuity of coal preparation operations into account.

Our country has scarce water resources, only equivalent to 25% of the world's per capita, and regional distribution is uneven. The differences between North-South are larger. At the same time, the main coal-producing areas have less water reserves. For example, water resources of Shanxi, Gansu and Ningxia (autonomous regions) accounted for only 7.5 percent of the northern water resources and groundwater natural resources accounted for only 8.9% of the northern groundwater natural resources [5] . So, the dry coal preparation technology is more adapted to China's national conditions, not only for the promotion of industrialization, but also for the on-site processing of coal conversion.

Since the fluidization technology has large capacity and the advantages of coupling reaction and is convenient to transfer material, the coal upgrading devices are in fluidized bed as the main body, mainly including coal separation equipment and pyrolysis equipment. To further meet the coal upgrading operations, recycling different grade waste heat of sintering and coking processes will be useful for the pyrolysis of coal and dry.

## 4.2. Mention Recycling Process

Above all, a new kind of coupled low-quality coal upgrading process was presented (Figure 4). Based on the control for the low-quality coal sorting, grading, controlled drying and pyrolysis processes, the component content of the low-quality coal (such as low-grade bituminous coal and lignite) was done to achieve the sintering and coking targets. Upon completion of pulverized coal recovery, the goal of the low-quality coal replacing part of the high-quality ones was also achieved. It not only recycled the waste heat of the systems, but also improved the utilization rate of coal.

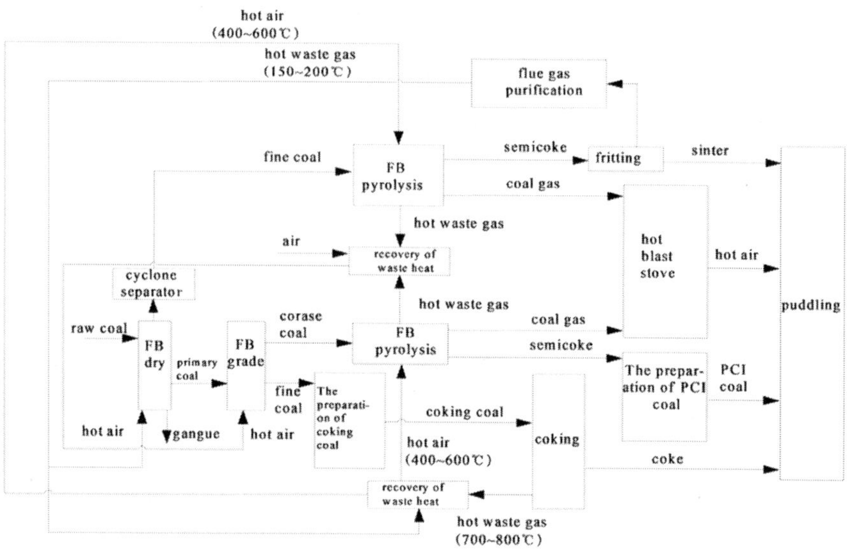


**Figure 4.** The system flow chart of low-quality coal upgrading.

## ACKNOWLEDGEMENTS

Bin Zhao, Qiang Yao, Shan Lu, Qinggong Wang, Junfu Lv (No. 2012CB214904) is sincerely acknowledged.

## REFERENCES

1. Li, Z., Yu, W., Yang, C. and Zhou, A.N. (2013) Actuality and Expectation on Upgrading Utilization of Low-Rank Coal. Mining & Processing Equipment, 41, 1-6.

2. Gu, X.-Y. (2009) Study on the Upgrading Process of Low-Rank Coal by the Thermalmodification. Clean Coal Technology, 1, 89-92.

3. Wang, J.-G. and Zhao, X.-H. (2012) Demonstration of Key Technologies for Clean and Efficient Utilization of Low-Rank Coal. Bulletin of the Chinese Academy of Sciences, 27, 382-388.

4. Guan, J., He, D.-M. and Zhang, Q.-M. (2011) The Technology of Improving Lignite Quality through Pyrolysis and the Concept of Poly-Generation. Coal Chemical Industry, 39, 1-4.

5. Qian, W. (2012) Experimental Study on Medium-Low Temperature Pyrolysis of Low Rank Bituminous Coal and Characterization of Pyrolysis—Derived Products. Ph.D. Thesis, China University of Mining and Technology, Beijing.

# Chapter 8

# Distribution of Mineral Species in Different Coal Seams of Talcher Coalfield and Its Transformation Behavior at Varying Temperatures

**Amit Banerjee[1] • P. R. Mishra[2] • Ashok Mohanty[1] • K. Chakravarty[3] • R. Das Biswas[1] • R. Sahu[2] • S. Chakravarty[1]**

*Analytical Chemistry Division, CSIR – National Metallurgical Laboratory, Jamshedpur, Jharkhand 831007, India*

## ABSTRACT

Mineral phase characterization and thorough understanding of its transformation behavior during combustion are imperative to know the potential utilization of coal in the thermal industries. The primary objective of this work is to analyze the quality of Indian Coals and obtain their mineral species-specific information at different depths. The samples were obtained from Talcher Coalfield, Odisha, India. Coal from four seam sections in the Talcher coalfield, India are mainly high ash coal (>50 %) and volatile matter deceases along with the seam depth. XRD results show that the major mineral phases present in the coal are quartz and kaolinite. Siderite, illite, and anatase were found in minor quantities. It has been observed that the clay minerals (kaolinite, silimanite, illite) decompose at higher temperature and traces of dolomite, mullite, hematite etc. are formed during the process of combustion. Among the four seams (M2,

M12, M24 and M43) studied, ash of M43 has high $Al_2O_3$%, $TiO_2$% and $K_2O$% content and low $SiO_2$%, CaO% and MgO% content. High acid-to-base ratios contributed to high ash fusion temperatures (IDT > 1500 °C) and low slagging potential of the coals studied. Relatively low fouling index (<0.3) was estimated for all the coal seams studied. Furthermore, thermodynamic modeling software, FactSage, have been used to envision the mineral phase transformations that take place between 800 and 1500 °C during coal combustion.

**Keywords:** Proximate analysis Ultimate analysis Fact Sage XRDAFT Talcher coalfield

## *1* INTRODUCTION

Power plants in India predominantly use coal as the energy source for power generation. Many problems associated with coal combustion, such as ash clinkering, slagging, fouling, agglomeration and deposition on heat exchange surfaces are routinely encountered in these power plants. Most thermal power industries generate database on elemental composition of coal ashes to predict the slagging and clinkering behavior of coal. However, only elemental compositions of coal ash cannot provide enough information on slagging characteristics of coal. Knowledge of structural characteristics of minerals and mineral phase transformation that occurs during the process of combustion of coal is essential to understand the slagging and clinkering behavior. There are a number of reports in the literature that provide information on the mineral matter association in coal (Van Dyk et al. 2009; Zhao et al. 2012; Dai et al. 2014a, b). Vassilev et al. (2005) have characterized mineral phase composition of composite samples from feed coals at Soma power station, Turkey. Mineral phases such as quartz, kaolinite, illite-muscovite, calcite, chlorite, plagioclase, pyrite etc. were identified in the feed coal at Soma power station, Turkey. Lopez and Ward (2008) have studied Combodian coal samples for mineralogy and kaolinite, illite, interstratified illite–smectite, quartz, and pyrite were the dominant mineral phases present in those coal samples. Recently, the process of phase transformation of mineral matter during coal combustion and gasification has been studied by some researchers. Chakravarty et al. (2015) have investigated the composition, mineral matter characteristics and ash fusion behavior of some Indian coals. FactSage thermodynamic model was used to understand ash fusion behavior and to predict the phase transformation that occurs during the process of coal combustion. Dai et al. (2014a, b) have studied origin of minerals and elements in the Late Permian coals, tonsteins, and host rocks of the Xinde Mine, Xuanwei, eastern Yunnan, China. Matjie et al. (2011) have investigated the mineral matter present in South African coal and their behavior in sintering and slagging of ash during the gasification process. The transformation behavior of mineral matters during combustion of Chinese coal gangue at varied temperature was studied by Zhou et al. (2012).

Indian Gondwana coals, in general, have high ash yields with low gross calorific value (Singh et al. 2013). In the commercial point of view, the coal quality in terms of fixed carbon content, composition and fusibility of coal ash are of prime importance. However, the fundamental concept of mineral matter transformation of high ash Indian coal during combustion requires a thorough investigation. The present study aims to investigate the mineral phase distribution at different seams of a particular borehole of Talcher coalfield, Odisha, India. Furthermore, the mineral phase transformation of coal at varying temperatures was also investigated experimentally and using the FactSage thermodynamic model.

# 2 MATERIALS AND METHODS

## 2.1 Sampling

Coal samples from a single borehole were collected from the Tentuloi Block, Talcher coalfield, Orissa, India. The coal bearing strata are made up of different coal seams as shown in Fig. 1. Coal samples used in the present study were from four different seams: seam No. M2 (depth: 154–155 m), M12 (depth: 241–296 m), M24 (depth: 337–341 m) and M43 (depth: 442–444 m). Coal samples were crushed and pulverized and passed through 212 microns mesh sieve size using a sieve shaker. The method of sampling and sample preparation for the chemical analysis was according to Indian Standard (IS: 436, 1964). Ashing of coal samples were done by heating at 750 °C in air for 1 h in a muffle furnace as per ASTM D3174-11. To study the mineral phase transformations at different temperatures, coal samples were heat treated at 600 and 1000 °C in air for 1 h in a muffle furnace.

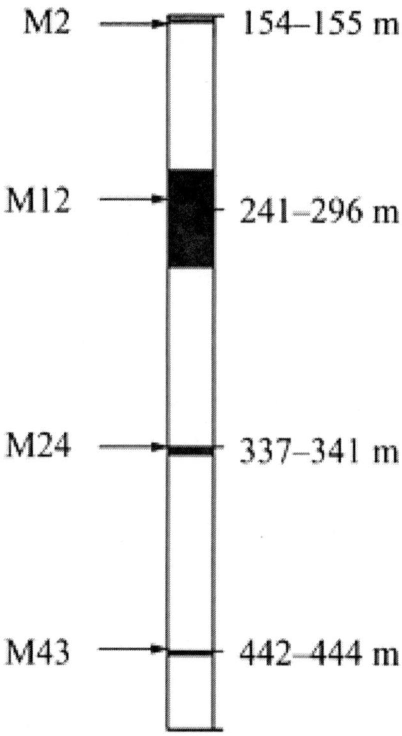


**Figure 1.** Stratigraphy of the Tentuloi Block of Talcher coalfield, Odisha, India and its coal seam nomenclature

## 2.2 Chemical characterization

Proximate analysis of coal on air dry basis was carried out using TGA-1000 instrument (NAVAS, USA) as per ASTM D7582-10 procedure. NIST coal SRM 1632d was used as the reference material. Ultimate analysis of coal was performed using CHNS analyzer, Vario EL-III (Elementar GmBH, Germany) as per ASTM D5373-08 procedure. The instrument was calibrated using NIST coal SRM 1632d reference material. The gross calorific value was determined using Parr 6200 bomb calorimeter (Parr, USA) following ASTM D5865-10a procedure. Benzoic acid was used as the standard to calibrate the bomb calorimeter. Chemical analysis of major inorganic elements present in coal and its ash samples were carried out using X-ray fluorescence (XRF) spectrometer (Bruker, S8 TIGER). For XRF analysis, sample preparation was done by press pellet method in which 2 g of wax micro-powder (Merck make) was mixed with 8 g of coal or coal ash sample and pressed at 30 ton pressure to make the pellets. QuantExpress semi quantitative analysis software (Bruker) was used for data analysis.

## 2.3 Mineralogical analysis

Mineral matter characterization of coal, heat treated coal and its ash samples were performed by X-ray powder diffraction (XRD) instrument (D8 Discover, Bruker, Germany). Ni-filtered CuKα was the radiation source and the scan was performed at $2\vartheta$ range of 10°–70° at a scan rate of 2°/min. HighScore Plus software package was used to identify mineral phases.

## 2.4 Ash fusion temperature

The ash fusion temperatures (AFT) of coal ashes were determined using AF 700 instrument (LECO, USA) under oxygen atmosphere in accordance with ASTM D1857 procedure. The procedure involves heating an ash cone up to 900 °C at 15 °C/min and then changing the heating rate to

5 °C/min from 900 to 1500 °C. During the process of AFT measurement, the initial deformation temperature (DT), softening temperature (ST), hemispherical temperature (HT) and fluid temperature (FT) were recorded according to the specific shape of the ash cone.

### 2.5 Thermodynamic calculations

Thermodynamic equilibrium calculation of coal ash in the oxidizing atmosphere was performed using FactSage software (version 6.4) to predict phase composition in the temperature range of 800–1500 °C at atmospheric pressure. Databases used for calculations were FactPS, ELEM, FTmisc, and FToxide. Elemental concentrations of coal ash were used as the input data for thermodynamic calculations.

## 3 RESULTS AND DISCUSSION

### 3.1 Coal quality and mineralogy

The quality of coal samples of 4 different seams M2, M12, M24 and M43 at varying depth of a particular borehole of Tentulai Block, Talcher coalfield were evaluated on the basis of the data obtained from proximate, ultimate and gross calorific value (GCV) analyses. The result (on air dry basis) is presented in Table 1. Coals of all the four seams are bituminous with reflectance of 0.51 and have ash yields in the range of 50.37 %–58.53 %, fixed carbon content in the range of 20.57 %–30.69 % and GCV in the range of 2810–3179 cal/g. Among the four seams studied, coal of M24 seam is of better quality having higher $FC_{ad}$ (30.69 %) and GCV (3179 cal/g). Coal at higher depth (M43 seam) was of relatively low quality having highest ash yields (58.53 %) and lowest GCV (2810 cal/g). Our earlier study also indicated that coal quality varied significantly from one seam to other for Indian coals (Chakravarty et al. 2015). It is also observed that $V_{ad}$ (%) and $S_{ad}$ (%) decrease with increase in depth.

**Table 1.** Proximate, ultimate and GCV analysis (air dry basis) of Talcher coals

| Seam | Proximate analysis (%) | | | Ultimate analysis (%) | | | | GCV (cal/g) |
|---|---|---|---|---|---|---|---|---|
| | $A_{ad}$ | $V_{ad}$ | $FC_{ad}$ | $C_{ad}$ | $H_{ad}$ | $N_{ad}$ | $S_{ad}$ | |
| M2 | 51.63 | 25.5 | 22.87 | 32.52 | 2.06 | 0.84 | 0.74 | 3114 |
| M12 | 50.37 | 24.9 | 24.73 | 33.51 | 2.18 | 0.96 | 0.64 | 3107 |
| M24 | 50.51 | 18.8 | 30.69 | 32.63 | 2.07 | 0.94 | 0.52 | 3179 |
| M43 | 58.53 | 20.9 | 20.57 | 26.86 | 1.68 | 0.51 | 0.30 | 2810 |

Mineral species characterization provides detail understanding of mineral phases present in coal and also indicates the probability of slag formation and behavior of ash fusibility during combustion (Matjie et al. 2011). Coal samples of different seams were studied for mineral phase characterization using X-ray powder diffraction technique. The XRD diffractogram of M2, M12, M24 and M43 coal samples (Fig. 2) indicates that quartz and kaolinite are the major minerals present in all the four coals. Minor amounts of illite, siderite and anatase minerals were also observed in higher depth of 442–444 m (M43). The XRD diffractogram for all the four coal samples are very similar, suggesting that same minerals are present in different coal seams of Tentulai Block, Talcher coalfield.

## 3.2 Chemical composition of coal ash

Chemical composition of coal ash is an indicator of mineral species present in coal samples and it can be also be used to predict slagging characteristics of coal. The chemical composition of ash samples for M2, M12, M24 and M43 coals was determined using X-ray fluorescence spectrometer. The data are presented in Table 2. $SiO_2$ and $Al_2O_3$ were the major constituents of coal ash, accounting to a total of 92 %–94 % of ash mass for all the four samples. The other constituents like trace amount of $Fe_2O_3$ (1.2 %–2.4 %), $TiO_2$ (1.3 %–2.0 %), CaO (0.2 %–1.3 %), MgO (0.3 %–0.7 %), $Na_2O$ (0.19 %–0.29 %) and $K_2O$ (0.38 %–0.52 %) were also present. Overall, the chemical composition of all the four ash

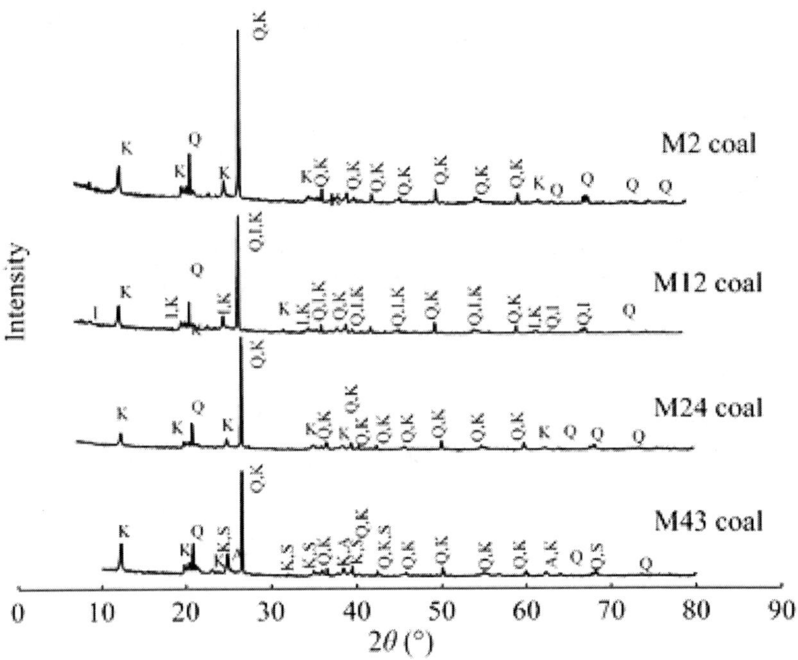


**Figure 2.** XRD diffractograms of M2, M12, M24 and M43 coal samples (Q—quartz, K—kaolinite, I—illite, A—anatase, S—siderite)

samples was similar with a slight variation of $SiO_2$ content (62.1 %–70.7 %). It is appeared from the Table 2 that the coal seam M43 is somewhat different from the other coal seam, M2, M12 and M24. The $SiO_2$%, CaO% and MgO% are lower whereas high concentrations of $Al_2O_3$, $K_2O$ and $TiO_2$ are present in coal seam M43 compared to the coal seam, M2, M12 and M24.

**Table 2.** Elemental composition of coal ash samples of Talcher coalfield (wt%)

| Seam | $SiO_2$ | $Al_2O_3$ | $Fe_2O_3$ | CaO | MgO | MnO | $TiO_2$ | $Na_2O$ | $K_2O$ |
|---|---|---|---|---|---|---|---|---|---|
| M2 | 69.6 | 22.6 | 1.8 | 1.3 | 0.7 | 0.01 | 1.3 | 0.22 | 0.45 |
| M12 | 66.9 | 24.8 | 2.4 | 1.0 | 0.5 | 0.02 | 1.5 | 0.26 | 0.38 |
| M24 | 70.7 | 23.6 | 1.2 | 0.2 | 0.5 | 0.01 | 1.3 | 0.29 | 0.39 |
| M43 | 62.1 | 31.2 | 2.4 | 0.2 | 0.3 | 0.03 | 2.0 | 0.19 | 0.52 |

The fouling property of coal ash is predicted from the chemical composition of ash by calculating fouling index using formula:

Foulingindex={(%Na2O+0.6589×%K2O)×%ash}/100

From the value of fouling index, the fouling property of ash is predicted. Ash having fouling index value <0.3 has low fouling property, 0.3–0.45 has medium fouling property, 0.45–0.6 has high fouling property and >0.6 has severe fouling property. In the present case, fouling index of M2, M12, M24, and M43 coal ash samples are calculated to be 0.24, 0.24, 0.26 and 0.30, respectively. Therefore, it is expected that the coal from seams of M2, M12 and M24 will have low fouling property whereas marginal fouling will occur for coal seams of M43.

### 3.3 Ash fusion temperatures

Ash fusion temperatures help in identifying the slagging behavior of coal ash, which affects slag viscosity and refractory life. The main constituents of coal ash used in the present study are silica and alumina with small percentages of other oxides such as $Fe_2O_3$, $TiO_2$, CaO, MgO, $K_2O$ and $Na_2O$. The ash fusibility is reported to be a function of these major oxides. Higher quantity of acidic compounds ($SiO_2$, $Al_2O_3$ and $TiO_2$) in coal ash contributes towards higher melting point. On the other hand, basic compounds in coal ash ($Fe_2O_3$, CaO and MgO, $Na_2O$ and $K_2O$) have a fluxing effect on $SiO_2$ and $Al_2O_3$, thereby reducing the fusion point of ash. Liu et al. (2013) have reported that three major chemical compositions that affect the AFTs of coal samples are CaO, and $Fe_2O_3$ content and $SiO_2$/$Al_2O_3$ ratio (S/A). AFTs decreases with increase of CaO and $Fe_2O_3$ content and increase of S/A ratio till 1.5. Beyond the S/A
...e of 2.0, AFTs decrease marginally or remain unchanged (Liu et al.
...he present case, the S/A ratio of three samples are higher
...0 and 2.99 for M2, M12, M24) respectively, and for
... Therefore the AFTs for all the coal samples
...ore, $Fe_2O_3$ and CaO content of all the
...ssibility of lowering of AFTs.

Indeed, the AFTs of M2, M12, M24 and M43 coal ash samples determined experimentally using ash fusion determination instrument was found to be very high. The initial deformation temperature (IDT) for all the four ash samples were higher than 1500 °C. The high AFT values are due to presence of high concentration of silica and alumina and appreciable amount of $TiO_2$ whose melting temperatures range between 1685 and 1775 °C (Borio and Levasseur 1984). Therefore, coal of the four seams of Tentulai Block, Talcher coalfield is good for combustion with very little possibility of ash fusibility at fuel bed of the boiler.

## 3.4 Phase transformation of mineral matter at high temperatures

Mineral matter present in coal undergoes phase transformations at high temperatures during combustion. To understand the process of phase transformations, coal samples M2, M12, M24 and M43 were heated at two different temperatures of 600 and 1000 °C for 1 h each in a muffle furnace in air atmosphere. XRD analysis of the heat treated samples was then carried out to identify the mineral phases formed after heat treatment. Figures 3, 4, 5 and 6 represents XRD of heat treated coal samples of M2, M12, M24 and M43, respectively at two different temperatures. At higher temperature of 600 and 1000 °C, a number of new phases are observed indicating that phase transformations of mineral matter have occurred during the heat treatment process. The kaolinite phase disappears and a new phase silimanite is formed at 600 °C. The silimanite phase is further transformed into cristobalite and mullite at 1000 °C as shown in Figs. 3, 4 and 5. Additionally, some new phases like hematite, dolomite, rutile $AlPO_4$ and feldspar have also appeared for M24, M43 samples as observed in Figs. 5 and 6. HighScore Plus software was further utilized to quantify the mineral phases present in heat treated coal samples. The data for M2, M12, M24 and M43 coal samples are tabulated in Tables 3, 4, 5 and 6, respectively. Quantitative estimation of mineral phases shows that all the clay minerals and aluminosilicates disappeared at high

(1000 °C), leaving behind $SiO_2$ (78 %–84 %) as the major phase along with other trace minerals as shown in Tables 3, 4, 5 and 6. Similar phase transformations have been reported in the literature (Pan et al. 2000; Bhargava et al. 2009).

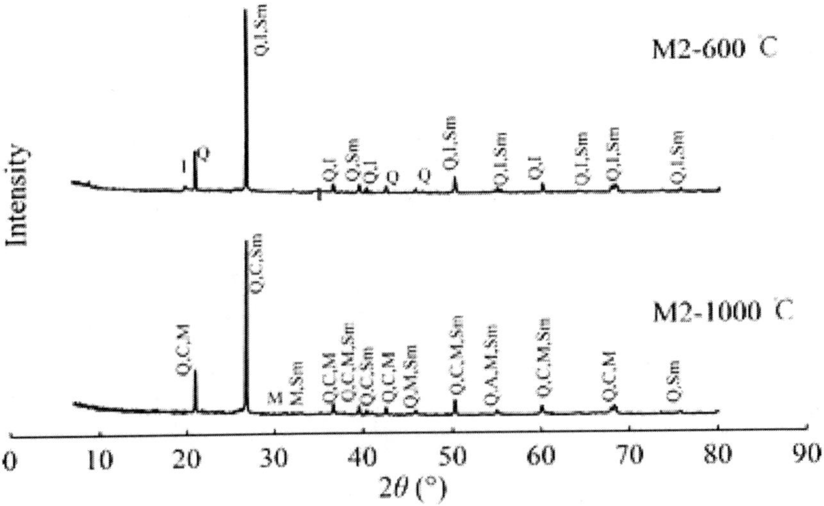


**Figure 3 .** XRD diffractograms of M2 coal samples heat treated at 600 and 1000 °C (Q—quartz, K—kaolinite, I—illite, A—$AlPO_4$, Sm—siliminite, C—cristobalite, M—mullite)

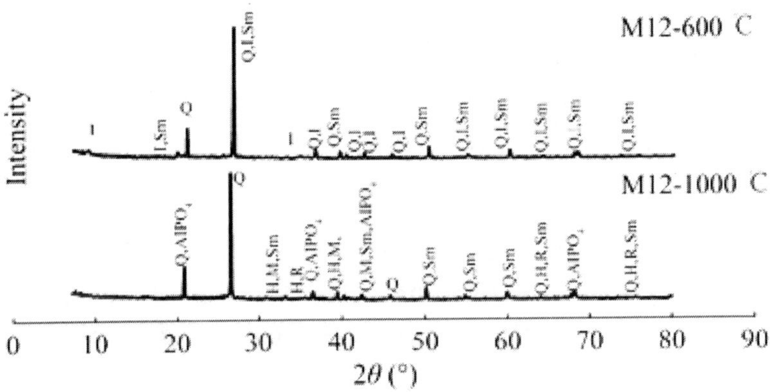


**Figure 4.** XRD diffractograms of M12 coal samples heat treated at 600 and 1000 °C (Q—quartz, K—kaolinite, I—illite, Sm—siliminite, C—cristobalite, M—mullite, H—hematite)

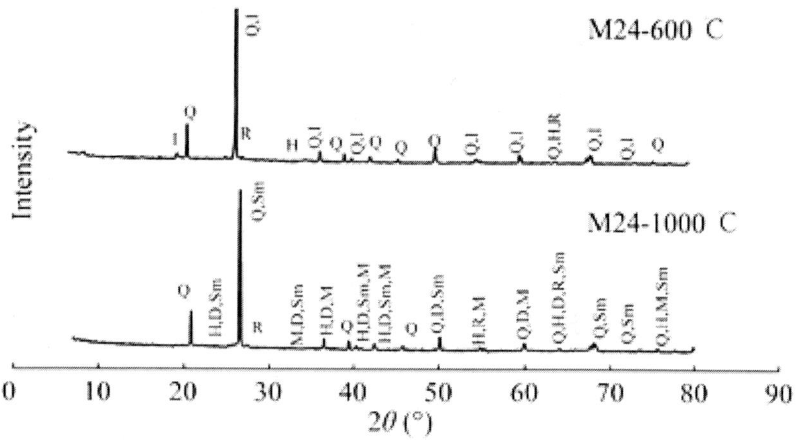


**Figure 5.** XRD diffractograms of M24 coal samples heat treated at 600 and 1000 °C (Q—quartz, K—kaolinite, I—illite, Sm—siliminite, C—cristobalite, M—mullite, H—hematite, R—rutile, D—dolomite)

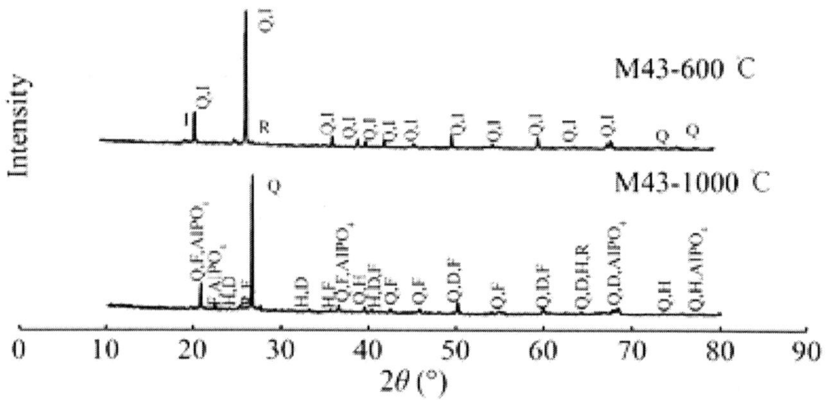


**Figure 6.** XRD diffractograms of M43 coal samples heat treated at 600 and 1000 °C (Q—quartz, K—kaolinite, I—illite, Sm—siliminite, C—cristobalite, M—mullite, H—hematite, R—rutile, D—dolomite, F—feldspar)

**Table 3.** Quantification of mineral species present in heat treated M2 coal samples (%)

| Minerals | 600 °C | 1000 °C |
|---|---|---|
| Quartz | 43 | 78 |
| Kaolinite | – | – |
| Illite | 49 | – |
| Silimanite | 8 | 8 |
| Cristobalite | – | 3 |
| Mullite | – | 5 |
| $AlPO_4$ | – | 6 |

**Table 4.** Quantification of mineral species present in heat treated M12 coal samples (%)

| Minerals | 600 °C | 1000 °C |
|---|---|---|
| Quartz | 41 | 82 |
| Kaolinite | – | – |
| Illite | 54 | – |
| Silimanite | 5 | 6 |
| Hematite | – | 2 |
| Rutile | – | 1 |
| Mullite | – | 5 |
| $AlPO_4$ | – | 4 |

**Table 5.** Quantification of mineral species present in heat treated M24 coal samples (%)

| Minerals | 600 °C | 1000 °C |
|---|---|---|
| Quartz | 57 | 84 |
| Kaolinite | – | – |
| Illite | 42 | – |
| Silimanite | – | 7 |
| Hematite | – | 3 |
| Rutile | 1 | 1 |
| Mullite | – | 3 |
| Dolomite | – | 3 |

**Table 6.** Quantification of mineral species present in heat treated M43 coal samples (%)

| Minerals | 600 °C | 1000 °C |
|---|---|---|
| Quartz | 34 | 78 |
| Kaolinite | – | – |
| Illite | 65 | – |
| Lizardite | – | – |
| Hematite | – | 3 |
| Feldspar | 1 | 10 |
| $AlPO_4$ | – | 4 |
| Dolomite | – | 3 |

### 3.5 FactSage thermodynamic modeling

FactSage 6.4 thermochemical software and databases computer package was used in order to predict (1) the chemical composition of coal ash at equilibrium during combustion (2) simulation of mineral phase transformation from 800 to 1500 °C, and (3) the relationship between mineral phases and ash fusibility. Chemical analysis data of ash samples was used as the input for the FactSage calculations.

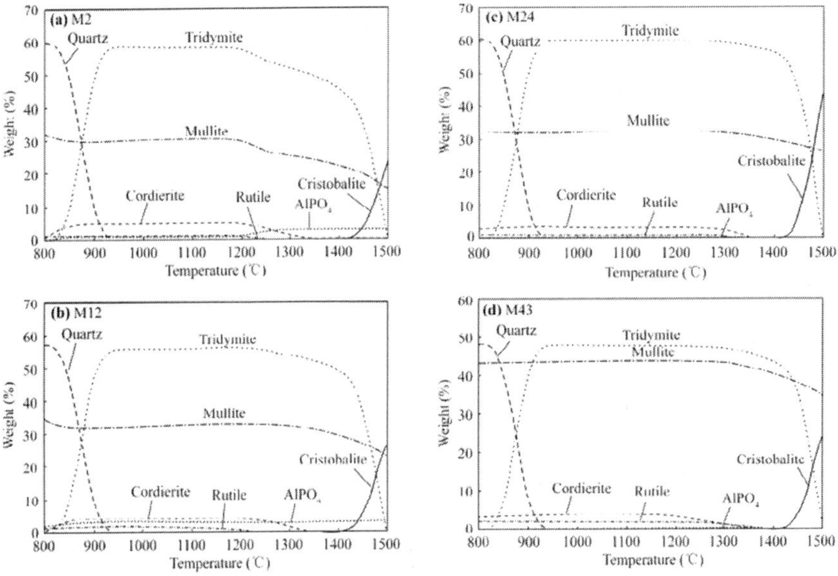


**Figure 7.** Phase transformation of coal ash sample obtained using FactSage

FactSage was first used to predict the mineral phase transformations that might occur at equilibrated conditions from 800 to 1500 °C for M2, M12, M24 and M43 coal ash samples. The results are presented in Fig. 7. Figure 7a shows that quartz and mullite will be predominantly formed at 800 °C in M2 coal ash sample. The major phase quartz remains till 850 °C, thereafter undergoes polymorphism and forms tridymite at higher temperatures. Aluminium phosphate ($AlPO_4$) and cristobalite ($SiO_2$) phase starts forming after 1100 and 1450 °C, respectively. The mineral phases predicted using FactSage agrees well with XRD result of M2 coal ash (Fig. 3). Figure 7b–d show a similar phase conversion trend for coal ashes of M12, M24 and M43, respectively. Rutile and cordierite present in low amounts for all the studied coal ash samples completely disappear after 1250 and 1350 °C, respectively. There are minor variations in composition between XRD and FactSage results as it is quite obvious that the FactSage predicts the stable mineral phases as a function of temperature whereas the mineral phases indicated by XRD analysis are the transient phases at varying temperature.

## 4 CONCLUSIONS

Investigations carried out on the coal samples of four different seams of a single borehole of the Tentuloi Block, Talcher coalfield, Orissa, India reveals that coals of all the four seams are similar in nature having high ash yields and low calorific values. The coal quality (FC content, GCV value) decreases marginally for the seam at higher depth. Mineral matter composition of the coal ash for the four coal seams were similar with $SiO_2$ and $Al_2O_3$ jointly accounting to a total of 92 %–94 % of ash mass. The coal ash composition in terms of major components, $SiO_2$, $Al_2O_3$, CaO, $TiO_2$, $K_2O$ and $Fe_2O_3$ however varied considerably between the seam M43 and other three seams, M2, M12 and M24. AFTs of coal ashes of all the four coal seams were very high with IDT greater than 1500 °C. Therefore, it indicates that coal of the four seams of Tentulai Block, Talcher coalfield is good for combustion with very little possibility of ash fusibility. Heat treatment studies of coal samples at different temperatures indicate that the mineral matter present in coal samples get transformed to different phases at different temperatures. New phases like hematite, rutile, mullite and $AlPO_4$ were formed at 1000 °C. Finally, FactSage thermodynamic modeling studies were carried to corroborate the findings of heat treatment experiments. FactSage predictions and experimental results agree well regarding phase transformation of mineral maters during heat treatment process.

## ACKNOWLEDGMENTS

The authors acknowledge and extend their sincere gratitude to the significant contribution and financial support provided by Dr. S. Srikanth, director of CSIR - National Metallurgical Laboratory, Jamshedpur, India.

## REFERENCES

1. Bhargava S, Garg A, Subasinghe N (2009) Insitu high-temperature phase transformation studies on pyrite. Fuel 88(1):988–993

2. Borio R, Levasseur A (1984) Overview of coal ash deposition in boilers. Energy Fuel 29(4):193–203

3. Chakravarty S, Mohanty A, Banerjee A, Tripathy R, Mandal GK, Basariya R, Sharma M (2015) Composition, mineral matter characteristics and ash fusion behavior of some Indian coal. Fuel 150:96–101

4. Dai S, Li T, Seredin VV, Ward CR, Hower JC, Zhou Y, Zhang M, Song X, Song W, Zhao C (2014a) Origin of minerals and elements in the Late Permian coals, tonsteins, and host rocks of the Xinde Mine, Xuanwei, eastern Yunnan, China. Int J Coal Geol 121:53–78

5. Dai S, Seredin VV, Ward CR, Jiang J, Hower JC, Song X, Jiang Y, Wang X, Gornostaeva T, Li X, Liu H, Zhao L, Zhao C (2014b) Composition and modes of occurrence of minerals and elements in coal combustion products derived from high-Ge coals. Int J Coal Geol 121:79–97

6. Liu B, He Q, Jiang Z, Xu R, Hu B (2013) Relationship between coal ash composition and ash fusion temperatures. Fuel 105:293–300

7. Lopez IC, Ward CR (2008) Composition and mode of occurrence of mineral matter in some Colombian coals. Int J Coal Geol 73(1):3–18

8. Matjie RH, French D, Ward CR, Pistorius PC, Li Z (2011) Behaviour of coal mineral matter in sintering and slagging of ash during the gasification process. Fuel Process Technol 92(8):1426–1433

9. Pan Y, Zhu RX, Banarjee SK, Gill J, Williams Q (2000) Papers on geomagnetism and paleomagnetism marine geology and geophysics-rock magnetic properties related to thermal treatment of siderite: behavior and interpretation. J Geophys Res 105(1):783–794

10. Singh PK, Singh GP, Singh MP, Naik AS (2013) Petrology of coals from Rampur Seam-IV and Lajkura seam, Ib River coalfield, Mahanadi Valley, Orissa, India. Energy Sources A Recovery Util Environ Eff Taylor Fr 35:1681–1690

11. Van Dyk JC, Benson SA, Laumb ML, Waonnders B (2009) Coal and coal ash characteristics to understand mineral transformations and slag formation. Fuel 88(1):1057–1063

12. Vassilev SV, Vassileva CG, Karayigit AI, Bulut Y, Alastuey A, Querol X (2005) Phase-mineral and chemical composition of composite samples from feed coals, bottom ashes and fly ashes at Soma Power Station Turkey. Int J Coal Geol 61(1):35–63

13. Zhao Y, Zhong J, Zheng C (2012) Transformation of aluminium rich minerals during combustion of a bauscite-bearing Chinese coal. Int J Coal Geol 94(1):182–190

14. Zhou C, Liu G, Yan Z, Fang T, Wang R (2012) Transformation behavior of mineral composition and trace elements during coal gangue combustion. Fuel 97(1):644–650

# Chapter 9

# Petrographic Characterization and Evolution of the Karharbari Coals, Talcher Coalfield, Orissa, India

**Alok K. Singh**[1]

*Rajiv Gandhi Institute of Petroleum Technology, Ratapur Chowk, Rae Bareli 229316, India*

## ABSTRACT

In the present study an attempt has been made to carry out the detailed petrographic characterization of the Karharbari coals of Talcher coalfield and to reconstruct the paleoenvironment conditions of coal formation using macerals and microlithtotypes as a tool. For these purposes a large number of samples were collected following the pillar sampling method and were subjected to detailed petrographic study. The petrographic observation shows that these coals are vitrinite rich followed by the liptinite and inertinite group of macerals. On microlithotype scale, these coals shows the dominance of the vitrite followed by clarite, vitrinertite and inertite. The concentration of liptite, clarodurite, duroclarite and vitrinertoliptite are insignificant. The vitrinite reflectance ranks the Karharbari coal as high volatile bituminous 'C' to high volatile 'B' bituminous. Coal petrography based depositional models suggest peat accumulation in forested telmatic swamp. Moreover, during the time of their evolution, there were alternate phases of oxic and anoxic moor conditions with good tissue preservation.

**Keywords:** Petrography Rank Evolution Karharbari Talcher coalfield

## 1 INTRODUCTION

The Gondwana coals of India are developed mainly along the two sides of a great triangular area whose third side is formed by the northern part of the east coast of Indian peninsula (Fig. 1a). Damodar, Son and Narmada Valleys make the north side of the triangle, whereas the southwestern side runs along to Godavari–Wardha Valley and subsidiary belt run along to Mahanadi Valley (Fig. 1a). The Talcher coalfield is the southeastern part of the Mahanadi Graben. It is detached basin surrounded by Precambrian rocks and occupies an area of over 1800 sq. km. in the Dhenkanal district of Orissa with a small segment lying in the adjoining Sambalpur district.

The first report on the occurrence of coal from this area dates back to 1937. The first systematic survey of the area was carried out by Blanford et al. (1856). Brief accounts of the geology, structure and coal seams of the basin have been given by Fox (1934), Pascoe (1959), Subramanian (1971), Das and Rath (1974) and Raja Rao (1982). Due to huge coal resources, the basin has attracted the attention of various exploring agencies. As a result the, the Indian Bureau of Mines, the Geological Survey of India, the Central Mine Planning and Design Institute, and the Directorate of Geology, Government of Orissa, initiated the exploration activities in the Talcher coalfield. In the year 1963–1965, Geological Survey of India conducted regional exploration in the coalfield. Presently the Talcher coalfield is exploited by Mahanadi Coalfield

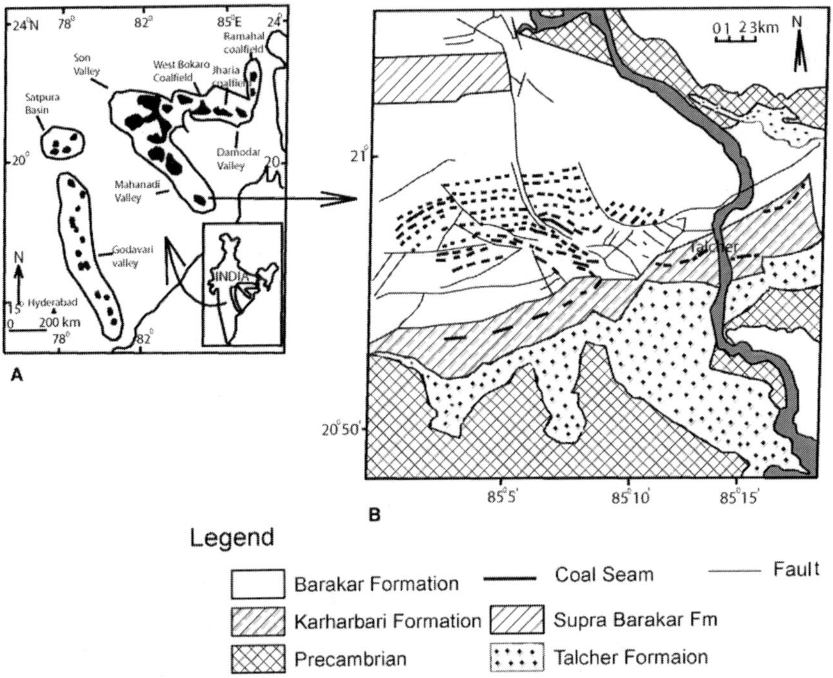


**Figure 1. a** Location map of Gondwana coal-fields of India. **b** Geological map of Talcher coalfield, Mahanadi valley, Orissa (Raja Rao 1982)

Limited, a subsidiary of Coal India Limited. Talcher basin is mainly studied by stratigraphers, sedimentologists and paleontologists, but petrological and geochemical aspects of Talcher coal, however, have not been taken into account so far. Although few reports have been published on the petrographic and geochemical characteristics of Talcher coals based on sporadic samples, the earlier studies lack the systematic sampling, regional approach and modern techniques. A mention may be the works of Das (1959), Navale (1965, 1966, 1971), Pareek (1955, 1956, 1958, 1963a, b), and Mishra et al. (1998) and Mohanty et al. (2001).

In the Talcher coalfield, coal seams have been reported in the Karharbari and the Barakar formation. In the present study an effort has been made to carry out a detailed petrographic and geochemical study of a large number of coal samples from the Karharbari Formation of the Talcher coalfield aiming to discuss the depositional conditions during accumulation.

## 2 GEOLOGY OF THE AREA

The rough outline of the coalfield resembles that of an ellipse. It is flat type of the basin and is bounded on all sides by Archeans. The Gondwana sediments of the investigated area represent a fairly continuous succession of strata comprising a Talcher and Damuda Groups. The sediment rest unconformably on the Precambrian basement comprising granites, gneisses, phyllites and amphibolites (Table 1).

**Table 1.** Stratigraphic sequence of Talcher coalfield (Raja Rao 1982)

| Age | Formation and lithology |
|---|---|
| Recent | Alluvium and laterite |
| Upper Permian–Triassic | Kamthi |
| | Fine to medium grained sandstone, carbonaceous shale, coal bands with greenish sandstone, pink clays and pebbly sandstones at top (250 m+) |
| Lower Permian | Barakar |
| | Medium to coarse grained sandstones, shales, coal seams with oligomictic conglomerate at base (500 m+) |
| Lower Permian | Karharbari |
| | Medium to coarse grained sandstones, shales and coal seam (270 m) |
| Lower Permian | Diamictite, fine to medium grained greenish sandstones, shale, rhythmite, turbidite etc. (170 m+) |
| Unconformity | |
| Precambrian | Granite, gneisses, amphibolites, migmatites etc. |

The Precambrian comprises a low ridge and isolated hillock along the boundaries of the basin. In Talcher coalfield, base of Gondwana sediments is marked by a pile of glacial and preglacial deposits. They were originally recognized by Blanford et al. (1856) and have been named Talchir Formation after the type locality. These rocks are well exposed along the southern boundary of the basin. This formation comprises diamictites, sandstone, needle shale, turbidites, rhythmites and varves (Table 1). The Talchir Formation conformably overlain by the Karharbari Formation, which is exposed as a narrow strip both along the southern and northern margins of the basin. Here, the Karharbari Formation has distinct lithological and palynological entity. This formation mainly consists of medium to coarse grained sandstone and shale. It also contains the superior quality of coal seam (Fig. 1b). Karharbari Formation is overlain by Barakar Formation, which characterized by a thick and conspicuous conglomerate horizon at the base. The conglomerate is oligomictic in nature and contains well round quartz and quartzite clast of various sizes. The basal conglomerate unit is overlain by a thick sequence (more than 500 m thick.) medium to coarse grained greyish feldspathic sandstone, grey to dark grey shale. The Barakar Formation conformably overlain by a sequence characterized by fine to medium grained, light grey and ferruginous sandstones and shale at the base and a thick succession of pale greenish sandstone with rare shale and pink clay bands, ferruginous coarse grained and pebbly sandstone at the top (Table 1). The formation is designated as Supra Barakar Formation.

## 3 METHOD OF STUDY

During the present study, coal samples were collected from Derra underground coal mines following the pillar coal sampling method. Coal samples were collected from Karharbari Top seam and Karharbari Bottom seam. Their megascopic characterization was performed by using the scheme proposed by Diessel (1965) and macroscopic seam profiles were constructed. For the microscopic study coal samples were crushed to 18 mesh size (<1 mm size particles). The polished particulate coal mounts were prepared by using cold mounting epoxy resin without pressure. The study was carried out on an advanced petrological microscope equipped with MSP 200 photometry system and fluorescence attachment. The maceral nomenclature applied, followed the ICCP system (1971, 1998, 2001). For the microlithotype analysis, a

20 point Kötter graticule was used. The line to line and point to point distance was maintained 0.4 mm for both maceral and microlithotype analysis. The maceral and microlithotype counting was done simultaneously. The reflectance measurement was carried out as per the ISO standard (ISO 7404-5 1994). For the precise assessment of liptinite and dark vitrinite microscopic examination of macerals was carried out both under the white incident light as well as under blue light excitation. The International Committees for coal Petrology (1971) devised certain rules for the characterization of maceral association. According to which, the minimum dimension of a band to be called a microlithotype, should be 50 × 50 mm, and the 5 % rule. The 5 % rule demonstrates that a microlithotype can have 5 % accessory macerals due to the very fact that neither the monomaceral nor the bimaceral microlithotypes are constituted exclusively of the macerals of one group or two groups respectively. Therefore, they can have 5 % accessory macerals which will have no role in their nomenclature or characterization. ICCP recommendation has been followed during the characterization of microlithotype. All the coal samples were subjected to proximate (BIS 2003) and petrographic analysis.

# 4 RESULTS

## 4.1 Macropetrographic characteristics of coal

These coals are banded in nature with vitrain and clarain being the dominant lithotypes. Durain has also been observed only in few coal samples and fusian has also been recorded. In some of the coal samples, pyrite framboids and small nodules have also been found. The scheme proposed by Diessel (1965) for the megascopic characterization of banded coals has been followed in the construction of megascopic seam profiles of Karharbari coals.

In the Karharbari formation, only bright coal, banded bright coal and banded coals have been recorded (Fig. 4).

## 4.2 Petrographic characteristics of coal

### 4.2.1 Rank

For the determination of rank of the Karharbari coals, the parameters used are volatile matter (d.a.f. basis) and vitrinite reflectance (Ro mean) value. In the Karharbari coals, the volatile matter (d.a.f. basis) ranges from 35.2 to 45.9 %, mean 43.1 % (Table 2) and reflectance of vitrinite (Ro mean) ranges from 0.57 % to 0.69 %, mean 0.59 % (Table 3; Fig. 4). The volatile matter and Ro mean values suggest these coals to range in rank from high volatile bituminous 'C' to 'B'.

**Table 2.** Proximate constituents (wt%) of Karharbari coals of Talcher coalfield, Orissa

| Proximate constituents | Top seam | | Bottom seam | |
|---|---|---|---|---|
| | Range | Mean | Range | Mean |
| Dry basis | | | | |
| Moisture | 7.0–8.0 | 7.9 | 8.0–10.5 | 9.0 |
| Ash | 10.0–20.0 | 13.7 | 5.0–11.3 | 8.7 |
| Volatile matter | 25.5–37.3 | 33.4 | 34.1–36.6 | 35.9 |
| Fixed carbon | 39.0–48.5 | 44.9 | 43.5–49.5 | 43.6 |
| Dry ash free basis | | | | |
| Volatile matter | 35.2–45.9 | 42.6 | 41.10–45.8 | 43.6 |
| Fixed carbon | 54.2–64.8 | 57.4 | 54.21–58.9 | 56.4 |

**Table 3.** Reflectance measurement of Karharbari coals of Talcher coalfield, Orissa

| Reflectance values | Top seam | | Bottom seam | |
|---|---|---|---|---|
| | Range | Mean | Range | Mean |
| Minimum | 0.39–0.46 | 0.4 | 0.33–0.44 | 0.37 |
| Mean | 0.57–0.69 | 0.59 | 0.57–0.61 | 0.59 |
| Maximum | 0.62–0.92 | 0.81 | 0.66–0.85 | 0.78 |
| Standard deviation | 0.06–0.11 | 0.08 | 0.06–0.09 | 0.08 |

### 4.2.2 Maceral composition

All the three group macerals viz. vitrinite, liptinite and inertinite have been recorded. Among the three macerals groups, vitrinite is the most dominant. The macerals of liptinite group have also occurred at moderate concentration, whereas macerals of inertinite group are low in concentration as compared to others two groups.

#### 4.2.2.1 Vitrinite group

All the maceral and sub-macerals of vitrinite group have been recorded in these coals. Telovitrinite in general, is the dominant subgroup. Among the telovitrinite subgroup, the concentration of telinite is quite poor while collotelinite occurs at high concentration and dominating over all the macerals of the vitrinite group. Generally it occurs as a thick band (Fig. 2b, c, h). The colour of collotelinite varies from light grey to dark grey. The cavities or cell lumens of the telinite is either filled with collotelinite or mineral matter. After the telovitrinite, the next dominating subgroup is detrovitrinite and represented by vitrodetrinite and collodetrinite (Fig. 2d, h). In the Karharbari coal beyond the collotelinite, next dominating maceral of the vitrinite group is vitrodetrinite. If vitrinite occurs as ground mass (mottled in appearance) of clarite and trimacerite with a colour darker and reflectance lower than the normal collotelinite in the same coal then such maceral is known collodetrinite. Gelovitrinite subgroup is represented by corpogelinite and gelinite. Corpogelinite is characterized by its typical oval to subrounded shape. In occurs occurs mostly as oval bodies as large isolated grains (Fig. 2a), in the form of cluster and also as cell fillings arranged in a definite pattern. The colour has been found to be paler and the reflectance is higher than the associated collotelinite. Gelinite occur as cell and fissure fillings. The colour of this maceral is slightly brighter and reflectance is comparatively higher than that of other vitrinite macerals in the same coal. Pseudovitrinite (more commonly called semi-vitrinite in India) is also present in these coals being characterized by its higher reflectance as compared to the normal collotelinite in the same coal and considered to be a transition stage between vitrinite and semi-fusinite. Some of vitrinite-bands of these coals have shown the fluorescence properties which are low reflecting and comparatively darker grey in colour. Such vitrinite is known as dark vitrinite. The fluorescence of vitrinite has been understood to be due to the presence of adsorped petroleum like substance, generated during the coalification process from liptinite and lipoid substance occurring in association of vitrinite (Teichmüller 1982a, b; Teichmuller and Durand 1983). Oxidation cracks are also recorded in few coals samples and cracks are in places filled with mineral matter (either argillaceous or sulphides). The quantitative distribution of vitrinite group of macerals of the Karharbari coals is given in the Table 4.

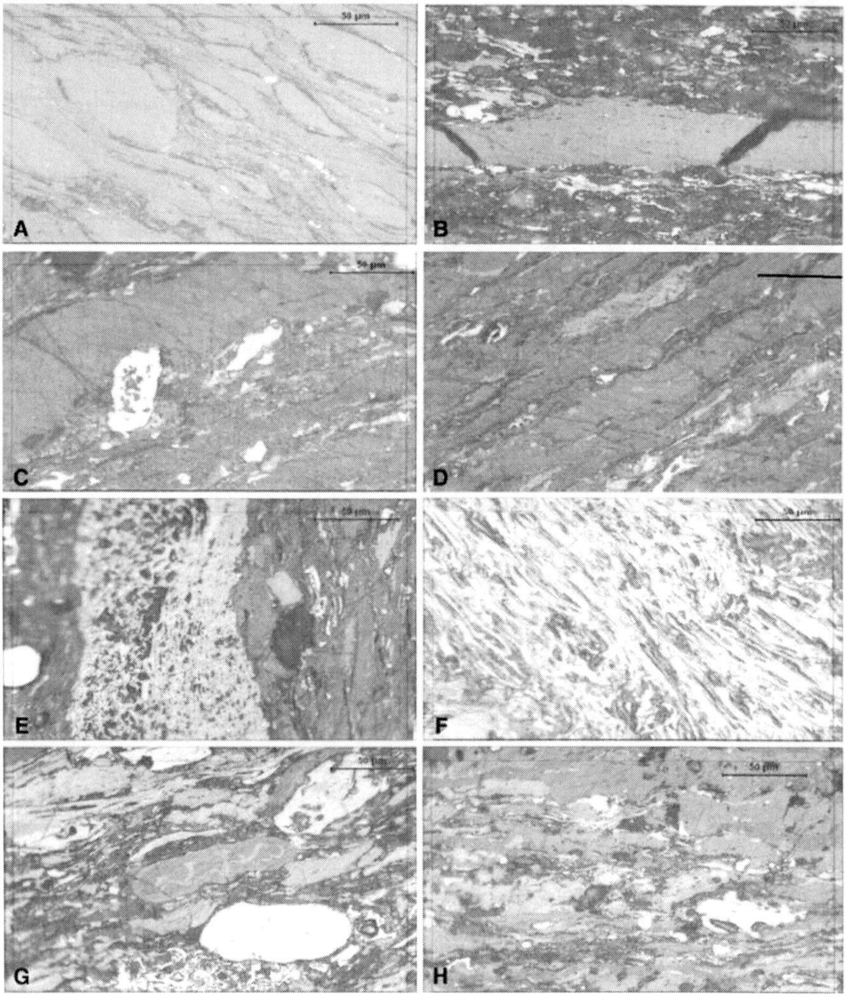

**Figure 2.** Photomicrograph under white incident light. **a** Carpogelinite, **b** band of collotelinite, inertodetrinite and carbonate mineral matter, **c** thick band of collotelinite and secretinite, **d** sporinite in the ground mass of collotelinite, **e** collotelinite, semifusinite, resinite and inertodetrinite, **f** fusinite, **g** macrinite, funfinite, **h** collotelinite, vitrodetrinite, inertodetrinite, argillaceous and carbonate mineral matter

**Table 4.** Maceral and mineral matter composition of Karharbari coals of Talcher coalfield, Orissa

| Maceral and mineral matter | Top seam | | Bottom seam | |
|---|---|---|---|---|
| | Range | Mean | Range | Mean |
| Vitrinite | 30.1–51.4 (43.9–62.6)* | 39.8 (53.8) | 31.6–55.8 (52.4–69.5) | 45.5 (61.9) |
| Telinite | Nil–1.5 (Nil–1.8) | 0.33 (0.4) | Nil–2.3 (Nil–3.6) | 0.62 (0.9) |
| Collotelinite | 15.2–44.9 (25.0–54.8) | 28.4 (37.8) | 26.2–51.0 (48.4–63.4) | 40.5 (55.1) |
| Vitrodetrinite | 1.5–13.4 (1.83–22.1) | 6.2 (8.8) | 0.31–1.5(0.4–2.4) | 1.07 (1.4) |
| Collodetrinite | Nil–1.4 (Nil–1.0) | 0.59 (0.8) | 0.3–1.2 (0.4–1.4) | 0.6 (0.8) |
| Corpogelinite | 0.4–2.3 (0.6–2.8) | 1.14 (1.6) | 0.4–1.7 (0.4–1.9) | 0.79 (1.0) |
| Gelinite | Nil–0.8 (Nil–1.0) | 0.4 (0.5) | Nil–0.8 (Nil–1.2) | 0.3 (0.4) |
| Dark vitrinite | Nil–2.3 (Nil–3.0) | 0.6 (0.8) | 0.4–1.0 (0.6–1.3) | 0.7 (0.9) |
| Pseudovitrinite | 0.8–4.6 (1.0–5.5) | 2.1 (2.8) | 0.4–2.1 (0.4–2.4) | 1.2 (1.5) |
| Liptinite | 17.9–29.3 (26.7–43.2) | 23.5 (31.9) | 11.6–29.1 (16.2–35.1) | 22.2 (27.2) |
| Sporinite | 6.2–11.5 (8.2–15.0) | 8.7(11.9) | 2.9–11.2 (3.7–14) | 6.6 (8.3) |
| Cutinite | 0.5–2.5 (0.7–3.1) | 1.3 (1.7) | Nil–1.5 (Nil–1.2) | 0.7 (0.8) |
| Resinite | Nil–1.1 (Nil–1.5) | 0.5 (0.7) | Nil–0.2 (Nil–0.2) | 0.02 (0.02) |
| Alginite | Nil–0.6 (Nil–0.5) | 0.1 (0.2) | Nil–0.7 (Nil–0.9) | 0.2 (0.3) |
| Liptodetrinite | 1.4–5.4 (2.3–6.9) | 3.1 (4.1) | 1.0–3.5 (1.5–4.8) | 2.6 (3.2) |
| Fluorinite | 0.7–2.1 (0.8–2.6) | 1.1 (1.5) | 0.6–3.1 (0.9–4.0) | 1.6 (2.1) |
| Exsudatinite | 1.2–6.5 (1.7–11.9) | 3.8 (5.2) | 3.5–7.4 (3.9–9.6) | 5.2 (6.6) |
| Bituminite | 0.2–2.1 (0.3–2.6) | 0.9 (1.2) | 0.2–2.7 (0.3–3.1) | 1.2 (1.5) |
| Degraded pollen | 0.4–5.83(0.6–7.7) | 3.0 (4.1) | 1.5–5.2 (2.0–6.6) | 3.8 (4.8) |
| Pollen sac | Nil–1.8 (Nil–3.1) | 0.9 (1.3) | Nil–1.4 (Nil–1.8) | 0.5 (0.6) |
| Inertinite | 5.55–17.2 (6.8–20.7) | 10.6 (14.3) | 4.5–13.1(5.8–20.6) | 8.7 (11.1) |
| Micrinite | Nil–0.2 (Nil–0.3) | 0.02 (0.03) | Nil–0.2 (Nil–0.2) | 0.02 (0.03) |
| Macrinite | Nil–0.4 (Nil–0.4) | 0.1 (0.2) | Nil | Nil |
| Semifusinite | Nil–4.9 (Nil–6.4) | 2.5 (3.2) | 0.9–3.5 (1.2–4.8) | 2.4 (3.0) |
| Fusinite | 0.9–8.4 (1.4–10.1) | 3.6 (4.9) | 0.9–3.1 (1.2–4.9) | 2.38 (3.0) |
| Secretinite | Nil–2.8 (Nil–3.5) | 1.0 (1.4) | 0.7–4.0 (0.9–4.8) | 1.8 (2.2) |
| Inertodetrinite | 0.7–4.7 (0.9–8.4) | 3.28 (4.5) | Nil–5.4 (Nil–8.5) | 2.1 (2.8) |
| Mineral matter | 16.63–39.11 | 26.21 | 15.16–39.21 | 20.43 |
| Argillaceous mineral matter | 14.3–34.5 | 21.81 | 10.1–30.1 | 17.4 |
| Carbonate mineral matter | 1.4–6.4 | 3.1 | 0.5–3.9 | 1.8 |
| Sulphide mineral matter | Nil–4.2 | 1.3 | 0.7–3.1 | 1.3 |

* Value given in parenthesis is mineral matter free basis

### 4.2.2.2 Liptinite group

In the Karharbari coals both primary (structured) and secondary liptinite (unstructured) liptinite macerals have been observed. In these coals, primary liptinite macerals are represented by sporinite, cutinite, resinite, alginite and liptodetrinite whereas secondary liptinitic macerals are represented by fluorinite, exsudatinite and bituminite. Degraded pollen and pollen sacs have also been recorded in the Karharbari coals and counted separately. Among the liptinite group the most dominating maceral is sporinite. In the Karharbari coals, both microspores (microsporinite) and megaspores (megasporinite) have been identified

(Fig. 3a–c). The colour of the spores is dark grey to brown under white incident light (Fig. 3d) while in fluorescent mode, the colour of the spores is yellow, turmeric yellow, orange-brown and brown colour (Fig. 3a–d, h). In the megaspore suture lines are quite distinct. The colour of cutinite is similar to sporinite dark grey to grey, but in some places it is paler than the associated sporinite. Thin walled cutinite has been identified in these coals. Resinite is also observed in the Karharbari coals and generally being oval to round in shape (Fig. 3a). In some samples, a thick band of weathered resinite has also been recorded. Resinite is more common in the Karharbari top seam as compared to bottom seam. In few samples of the Karharbari coals, alginites is also identified (Fig. 3e). It indicates the marine incursion during the deposition of coals. The alginite can hardly be identified under white incident light. However, it can be better studied in blue irradiation where it shows green, yellow and yellowish- brown colours depending upon the rank of coal. Its fan shaped morphology indicates the pila type of alginite present in the Karharbari coals. Occurrence of acritarch and and ichnofossils have also indicated the marine signature in the basin (Tiwari et al. 1995; Goswami, 2002). Secondary liptinite macerals (unstructured liptinite) have identified at higher concentration indicating that these coals can act as a source rock for the hydrocarbon generation. To support these observations geochemical data are very much essential. Bituminite appears as amorphous, as well as lamellar forms without any definite shape and size occurring as finely dispersed lenses streaks and at places as groundmass for other liptinite maceral (Fig. 3e, h). It shows characteristic orange or brown colour under fluorescence light. Exsudatinite occurs as a crack or fissure fillings in collotelinite and shows black colour under white reflected light and s orange and brownish orange under blue irradiation fluorescence (Fig. 3g). Fluorinite is very common in these coals (Fig. 3h). It is seen as black colour, elongated or lensoid bodies of variable thickness under the white reflected light. Under the blue excitation, it is greenish yellow to yellow. Liptodetrinite is frequently occurring in these coals (Fig. 3a, d, f, g). The quantitative distribution of liptinite group of macerals of the Karharbari coals is given in the Table 4 and Fig. 4.

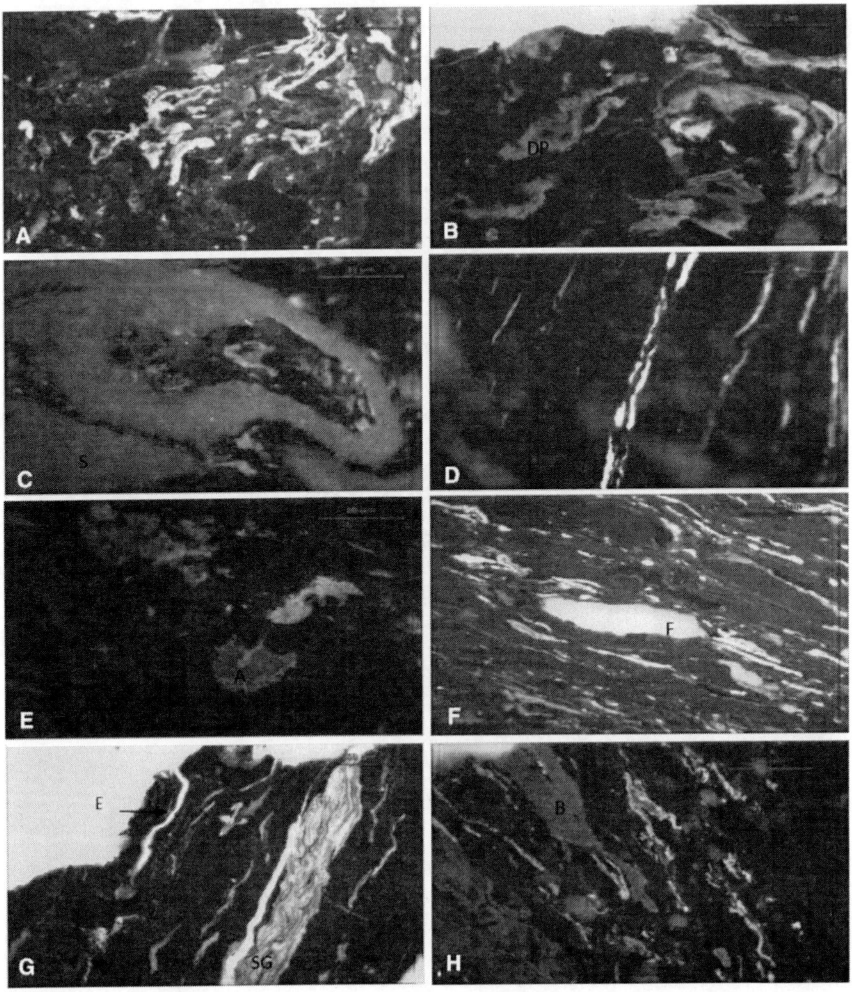


**Figure 3.** Photomicrograph of liptinite macerals under fluorescent light. **a** spores, small resin bodies and liptodetrinite, **b** sporinite and degraded pollen (DP), **c** megaspore (S), **d** fluorinite (F) and spore, **e** aliginite (A) and amorphous bituminite, **f** fluorinite, liptodetrinite and spores, **g** exsudatinite (E) and sporangium (SG), **h** amorphous bituminite (B), liptodetrinite and spores

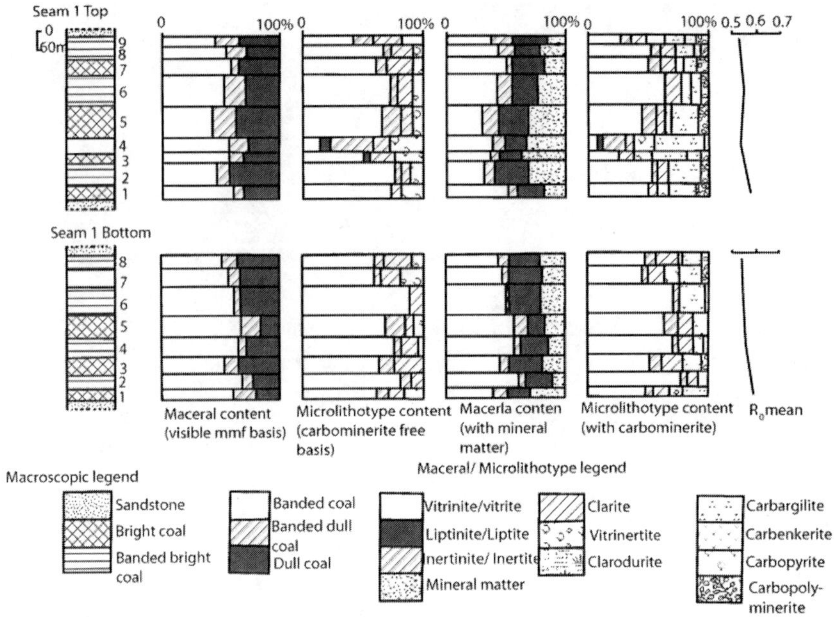


**Figure 4.** Macropetrographic section and microscopic seam (maceral and microlithotype) and range of mean vitrinite reflectance (in oil) of Karharbari coals Talcher coal field, Mahanadi valley, Orissa

### 4.2.2.3 Inertinite group

Inertinite macerals have the same precursors as vitrinite macerals and many of them pass through the same stages of humification except that, before reaching depositional base level below the groundwater table, they are subjected to a period of intensive desiccation and varying degrees of oxidation including partial burning of the accumulated vegetal matter (Gould and Shibaoka 1980). The results are coal constituents which possess relatively high O/C ratios and high reflectance in incident light microscopy because they are rich in aromatic carbon. Most inertinite macerals are relatively brittle and hard, which in incident light microscopy is shown by their tendency to develop polishing relief. Inertinite includes a group of maceral which are characterized by their white to yellowish white colour and high reflectance.

In the coal of Karharbari Formation inertinite occurs at low concentration as compared to the macerals of vitrinite and liptinite groups. All the macerals of this group occur in these coals with significant dominance of fusinite and semifusinite. The macerals of this group include micrinite, macrinite, semifusinite, fusinite, sclerotinite and inertodetrinite.

Fusinite is characterized by a well preserved cellular structure where cell lumens may be open or occupied by mineral matter (carbonate, pyrite and clay) (Fig. 2f). Bogen and sieve structures have also been observed in fusinite. Under white incident light the colour of fusinite is white to yellowish white with high relief and very high reflectance. Both pyrofusinite and degrado fusinite have been recorded. After the fusinite, next dominating maceral of the inertinite group is semifusinite (Fig. 2e). The cell structure of semifusinite is not so well preserved as in the case of fusinite. Its colour lies between light grey to white. In some places, transition from collotelinite to semifusinite has also been observed during the petrographic study. Figure 2e shows the sharp boundary between collotelinite and semifusinite. Marionette has also been observed in these cells. It occurs as rounded to oval bodies. (Figure 2g). The colour is white and reflectance is high close to this fusninte in the same coal. Micrinite has been observed in these coals and it occurs in the form of lenses. It shows white colour and high reflectance. Secretinite has also been recorded in these coals (Fig. 2c). Inertodetrinite is observed only in these coals and it appears to have been derived from crushing of fusinite and semifusinite (Fig. 2c, g, h). The quantitative distribution of inertinite group of macerals of the Karharbari coals is given in the Table 4 and Fig. 4.

### 4.2.3 Mineral matter

The minerals present in these coals includes clay, carbonate, and sulphides. The clay minerals occur as ground mass and also as discrete form, fissure, crack, and cleat fillings and as a infilling of cell lumens of telinite and mainly detrital in orgin. Sulfide occurs in the form of pyrite as disseminated pyrite, discrete grains and infillings of fissures and cracks. Pyrite particles of <1 mm size have been identified as disseminated or submicron pyrite. In the coals it occurs mostly as sporadically distributed particles in the organic ground mass. This form of pyrite originates during hummification process and thus characterized genetically as syngenetic (Reyes-Navarro and Davis 1976; Renton 1979). The discrete grain of pyrite is characterized as a epigenetic origin, which comes into being through ingress of ferruginous solutions at the later stages of coal formation, particularly after partial compaction (Neavel 1966). Fissure and crack filling form of the pyrites are very common in these coals and occurrence of these pyrite forms are described to be of epigenetic origin (Renton, 1979). In one or two samples, framboidal pyrite has also been observed. Although mechanism of the development of framboid is yet to be understood; Kizilstein and Trufanov (1968), Kizilstein and Minaeva (1972), Skripchenko and Berberian (1975), and Reyes-Navarro and Davis (1976) have shown that this type of pyrite is produced by bacteria in anaerobic settings and which may be considered as marine signature. The carbonates occur as ground mass, as strings and also as a cavity and fissure fillings. At few places it has impregnated over the macerals. Carbonate minerals are mainly syngenetic in origin. The concentration of different type of mineral matters in the Karharbari coals is given in the Table 4 and Fig. 4.

### 4.2.4 Microlithotype and carbominerite

Among microlithotype group, the most dominating is monomaceral followed by bimaceral and trimaceral. The microlithotype analysis shows the dominance of vitrite, clarite, vitrinertite and inertite. Monomaceral microlithotype group is represented by the vitrite, liptite and inertite whereas bimaceral microlithotype group is represented by the clarite and vitrinertite. Trimaceral group has been seen only a few samples. The mineral matter has also been identified in terms of carbominerites. The carbominerite has been characterized as carbargilite, carbankerite, carbopyrite and carbopolyminerite (Mackowsky 1982). The most dominating carbominerite in the Karharbari coals is the carbargillite and which followed by the carbankerite, carbopyrite and carbopolyminerite. The concentration range of the different microlithtypes is given in the Table 5 and Fig. 4.

**Table 5.** Microlithotype and carbominerite composition of Karharbari coals of Talcher coalfield, Orissa

| Microlithotype & carbominerite | Top Seam | | Bottom Seam | |
|---|---|---|---|---|
| | Range | Mean | Range | Mean |
| Monomaceral | | | | |
| Vitrite | 7.4–65.6 (41.7–75.8)* | 41.7 (59.5) | 43.7–76.5 (59.2–89.1) | 58.1 (70.7) |
| Liptite | Nil–3.7 (Nil–4.4) | 0.9 (1.6) | Nil–1.0 (Nil–1.3) | 0.3 (0.3) |
| Inertite | Nil–18.5 (Nil–16.7) | 7.5 (10.4) | 3.4–11.7 (2.3–13.6) | 6.7 (8.2) |
| Bimaceral | | | | |
| Clarite | 3.7–15.1 (5.5–26.1) | 9.3 (14.2) | 4.9–17.0 (6.3–21.6) | 10.7 (13.4) |
| Vitrinertite | 4.1–14.8 (5.3–28.6) | 8.6 (13.7) | 1.2–11.7 (1.6–15.8) | 5.1 (6.6) |
| Trimaceral | | | | |
| Duroclarite | Nil–1.5 (Nil–1.8) | 0.4 (0.6) | Nil–1.1 (Nil–1.4) | 0.4 (0.5) |
| Clarodurite | Nil | Nil | Nil–1.1 (Nil–1.4) | 0.2 (0.3) |
| Vitrinertoliptite | Nil | Nil | Nil–1.1 (Nil–1.4) | 0.2 (0.3) |
| Carbominerite | | | | |
| Carbargilite | 5.5–35.4 | 22.6 | 4.8–19.4 | 12.6 |
| Carbankerite | 1.4–10.4 | 3.6 | 0.4–1.9 | 1.2 |
| Carbopyrite | Nil–5.3 | 2.1 | 0.6–3.9 | 1.9 |
| Carbopolyminerite | Nil–10.5 | 4.3 | 0.8–7.7 | 2.7 |

* Value given in parenthesis is carbominerite free basis

## 5 DISCUSSIONS

Advancement of coal petrographic techniques offers extended parameters which can retrieve the archived records of the evolutionary path of coals. The qualitative and quantitative occurrence of microscopic codes which are to be decoded to interpret the nature and type of plant community, depositional milieu, dynamics of surface and underground waters, paleogeography and paleoclimate of a coal deposit. The coal petrologists who have significant contribution in this field are: Krausel (1961), Cohen and Spackman (1972), Smyth (1979, 1980, 1984); Teichmüller and Teichmüller (1975); Teichmuller and Teichmuller (1982); Styan and Bustin (1983), Harvey and Dillon (1985), Diessel (1986), Cohen et al. (1987), Hunt (1989), Hunt and Smyth (1989), Calder et al. (1991), Grady et al. (1993), Shearer and Clarkson (1998), Hawke et al. (1999), Duan et al. (2011), Deng and Sun (2011), Lin and Tian (2011), Singh and Singh (1996, 2000), Singh et al. (2010a, b, 2012a, b, 2013) and Suárez-Ruiz et al. (2012).

Coal facies is a function of the type of peat forming flora. The flora in a given geological milieu are controlled by climate, ground water table, nutrient supply and chemistry of water (marine influence). The ground water is the most important factor which determines not only the type of vegetation but also controls the redox potential (oxic to anoxic) and thereby the mode of preservation of plant litter or intensity of humification process. A peat accumulating system can be distinguished in coastal lowland and freshwater peat land (Diessel 1992). The coastal low land is protected by sand bars or barrier beaches, close to tidal influences and grade into marine-influenced to freshwater mires. The fresh water peatlands are inland mire, having no connection to sea, including upper delta and alluvial plain swamps, marshes and bogs. Peat may form in forest swamps from various plant associations, or in marshes with herbaceous vegetation (sedges and grasses) or in open swamps with predominantly submerged and floating plants or in raised bog with mosses or shrubs or trees. Several coal geologists made significant contributions towards the depositional system, mention may be the name of Amijaya and Littke (2005), Marques (2002), Sebag et al. (2006), and Silva et al. (2008). For retrieving the type of Karharbari coal facies, petrography based model proposed by Hacquebard and Donaldson (1969), Diessel (1986), Calder et al. (1991), Mukhopadhyay (1986), Singh and Singh (1996) and Singh et al. (2013) have been used.

The microlithotype composition of the Karharbari coal in a facies model proposed by Hacquebard and Donaldson (1969) suggests that this coals was evolved from the forest mire (Fig. 5). Hacquebard and Donaldson (1969) have shown that the quality and quantity of flora are controlled by water depth conditions of the peat mires, which also regulates the mode of preservation and thereby quantitative distribution of macerals and microlithotypes. Therefore, the microlithotype composition of Karharbari coal suggests the evolution of these coals via forest moors. In order to substantiate the above contemplation a facies model designed by Diessel (1986) has been tested for these coals. Diessel has formulated two parameters—gelification index (GI) and tissue preservation index (PI), for retrieving the paleofacies and has interpreted the depositional setting based on these GI and TPI index have been calculated according to the given formulae:

GI=Vitrinite + MacriniteSemifusinite + Fusinite + Inertodetrinite

TPI=Telinite + Collotelinite + Semifusinite + FusiniteCollodetrinite + Macr inite + Inertodetrinite

This model was further addressed by Lamberson et al. (1991), Alkonde et al. (1992) and Diessel (1992). The TPI is the measure of the degree of tissue decomposition and the ratio of the woody plants in the original peat forming plant congregation. Further, the ratio of tissue-derived structured macerals versus tissue-derived unstructured macerals is a scale to quantify the degree of humification suffered by maceral precursor and also to measure the proportion of wood matter contributed to the gamut of peat. In addition, low TPI gives two different clues either the dominance of herbaceous plants in the swamp or an intense humification process leading to extensive destruction of plant tissues (Diessel 1992). On the other hand, high TPI suggests the presence of well-preserved plant tissue (telinite, fusinite and semifusinite), a high proportion of arboreal vegetation (telinite, collotelinite, fusinite and semifusinite) and aerobic condition.

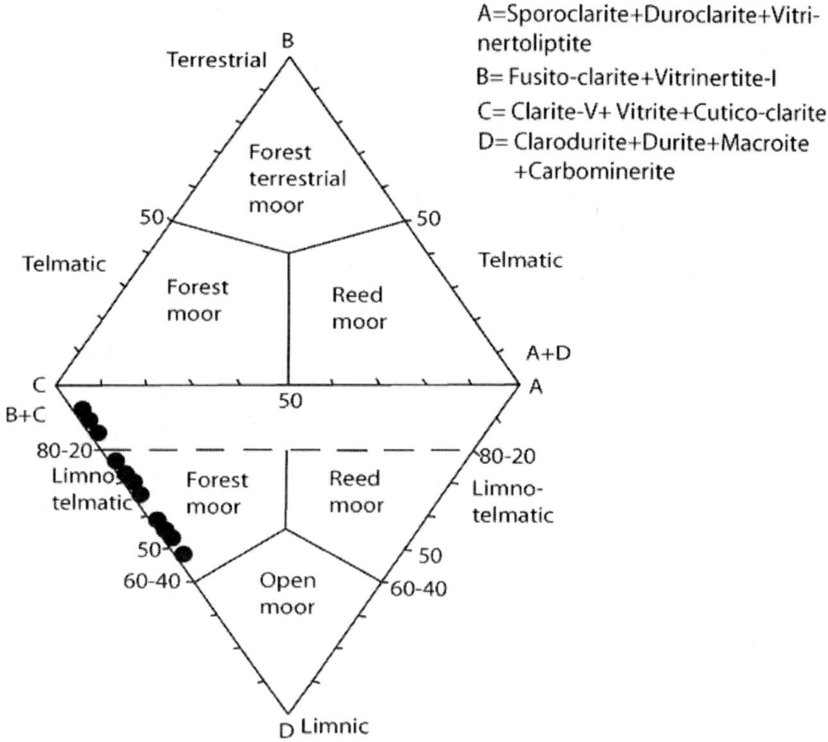


**Figure 5.**Microlithotype composition of Karharbari coals plotted on a facies diagram proposed by Hacquebard and Donaldson (1969) and modified by Marchionni (1980)

The GI scales the intensity and time of wet condition (Diessel 1992). It decreases with the increase of oxidation. Further, a high GI and TPI point to a meagre aerobic decomposition (Lamberson et al. 1991). In case of Karharbari coal, the GI and TPI values are moderately high. The GI and TPI values of Karharbari coal in this model (Fig. 6) indicates that these coals have evolved under telmatic conditions and wet forest swamp.

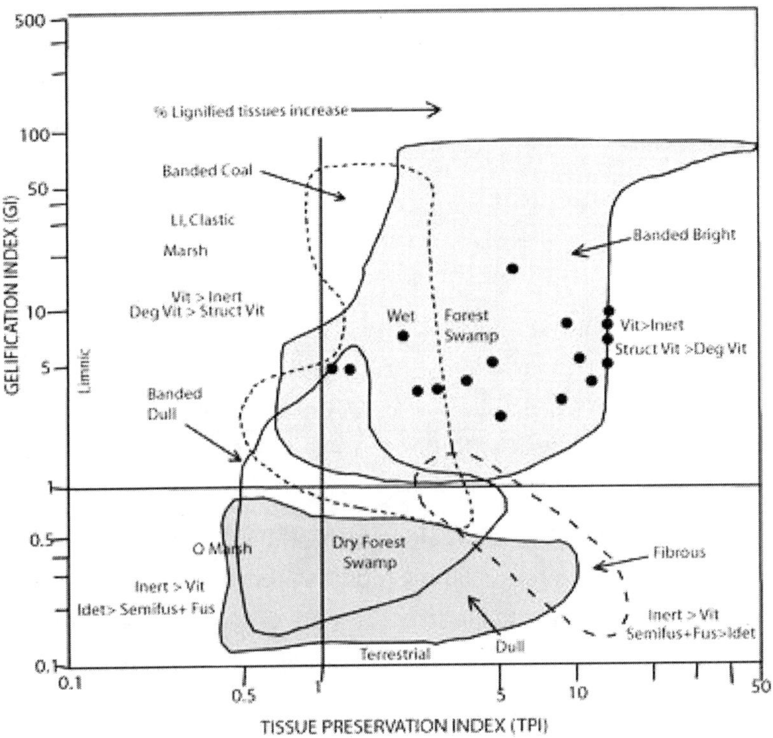


**Figure 6.** Coal facies deciphered from gelification index (GI) and tissue preparation index (TPI) in relation to depositional setting and type of mire for Karharbari coals, Mahanadi valley, Orissa (after Diessel 1986 and modified by Kalkreuth et al. 1991). *LI* limited influx, *O MARSH* open marsh, *VIT* vitrinite, *INERT* inertinite, *SEMIFUS* semifusinite, *FUS* fusinite, *IDET* inertoderinite, *STRUCT* structured, *DEG* degraded

A facies model of Calder et al. (1991) is also used, which is based on the ground water index (GWI) and vegetation index (VI). In fact, this model is very similar to GI and TPI indices of Diessel (1986). The parameters used in the reconstruction of peat lands are the degree of ground water influence, relative rainfall (Kalkreuth et al. 1991; Ligouis and Doubinger 1991), changes in ground water level in the mire (Peatland = marsh + swamp), vegetation, mineral matter content and degree of preservation of maceral precursor (Calder et al. 1991). The GWI and VI are calculated as under:

$$\mathrm{GWI} = \frac{\text{Gelinite} + \text{Corpogelinite} + \text{Clay mineral} + \text{quartz} + \text{Vitrodetrinite}}{\text{Telinite} + \text{Collotelinite} + \text{Collodterinite}}$$

$$\mathrm{VI} = \frac{\text{Telinite} + \text{Collotelinite} + \text{Fusinite} + \text{Semifusinite} + \text{Suberinite} + \text{Resinite}}{\text{Collodterinite} + \text{Inertodetrinite} + \text{Alginite} + \text{Liptodetrinite} + \text{Sporinite} + \text{Cutinite}}$$

$$\mathrm{GI} = \frac{\text{Vitrinite} + \text{Macrinite}}{\text{Semifusinite} + \text{Fusinite} + \text{Inertodetrinite}}$$

$$\mathrm{TPI} = \frac{\text{Telinite} + \text{Collotelinite} + \text{Semifusinite} + \text{Fusinite}}{\text{Collodetrinite} + \text{Macrinite} + \text{Inertodetrinite}}$$

Based on GWI and VI index, Calder et al. (1991) retrieved major mire paleoenvironments- such as limnic (open water marsh), swamp, fen and bog (Fig. 7) representing rheotrophic and ombrotrophic hydrological conditions. The GWI and VI values of Karharbari coals suggest the deposition of peat in swamp forest, mostly under rheotrophic to mesobrotrophic hydrological conditions, but few plots also lying in the ombrotrophic zone (Fig. 7).

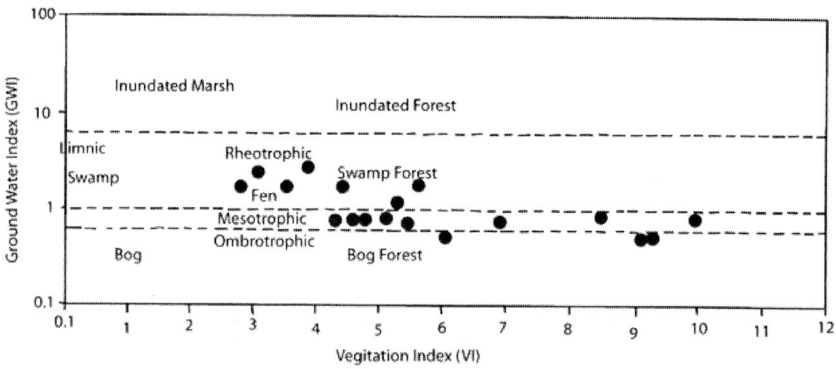


**Figure 7.**Karharbari coals, Talcher coalfield, Mahanadi valley, Orissa, GWI versus VI Paleoenvironment diagram (modified after Calder et al. 1991)

A facies model proposed by Mukhopadhyay (1986) has also been used for the Karharbari coals based on the maceral composition. In this model (Fig. 8), plots of Karharbari coals indicate again their origin by forest swamp under mildly oxic to anoxic conditions with good tissue preservation.

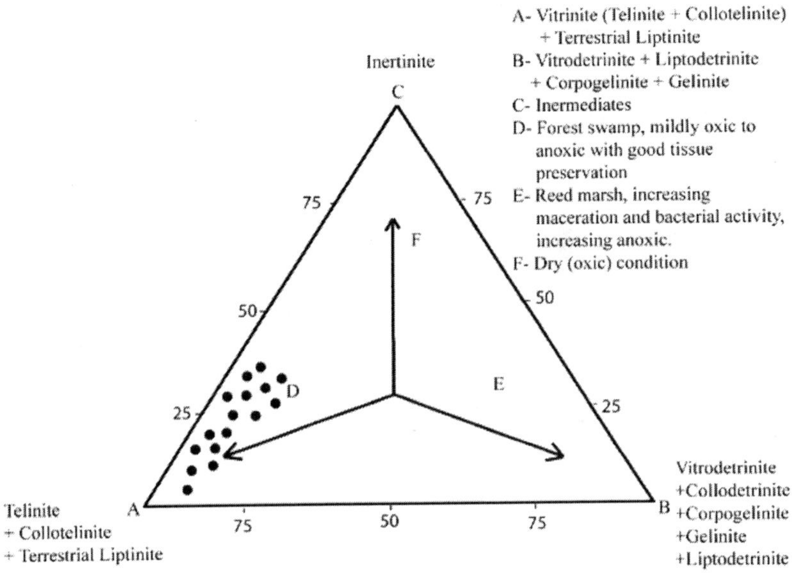


**Figure 8.** Ternary diagram illustrating facies critical maceral association in Karharbari coals, Talcher coalfields, Mahanadi valley, Orissa and suggested peat forming environment (modified from Mukhopadhyay 1986)

A facies model proposed by Singh and Singh (1996) has been applied which is based on the maceral and mineral matter content. In this model more emphasis has been given in the quantitative occurrence of mineral matter in coal because it directly relates to the influx of surface water cover in swamps. The plots of Karharbari coal in this model suggest that these coals was deposited in alternating oxic and anoxic mire (Fig. 9) It is indicative of fluctuating water cover in the basin.

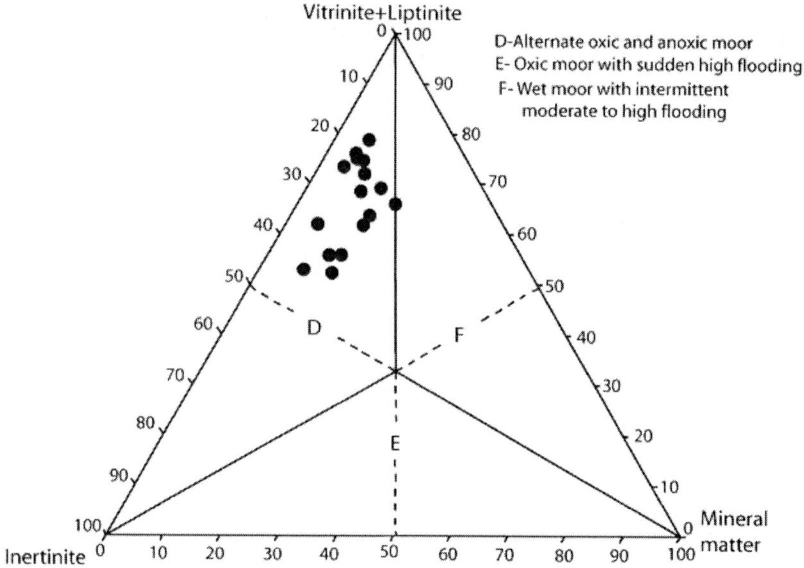


**Figure 9.** Depositional conditions of coals based on maceral and mineral matter content (after Singh and Singh 1996)

A = Liptite + Clarite (L) + Vitrinertoliptite + Durite (L)

B = Vitrite + Clarite(V) + Vitrinertite (V) + Duroclarite

C = Inertite + Durite (I)| + Clarodurite + Vitrinertite (I)

The types of facies delineated are:

CF1 Deep water facies, characterized by organic mud, rich in liptinites
CF2 Open water with dominance of subaquatic plants
CF3 Forest swamp with sudden eustatic changes
CF4 Forest to reed swamp
CF5 Forest swamp
CF6 Moss swamp with extremely dry conditions

Yet another facies model based on Microlithotype has also been used (Singh et al. 2013) to infer the paleodepositional conditions of coals. Three parameters (A, B and C) have been taken into considerations which are as under:

The plots of Karharbari coal, on this model, suggest the development of forest facies (Fig. 10).

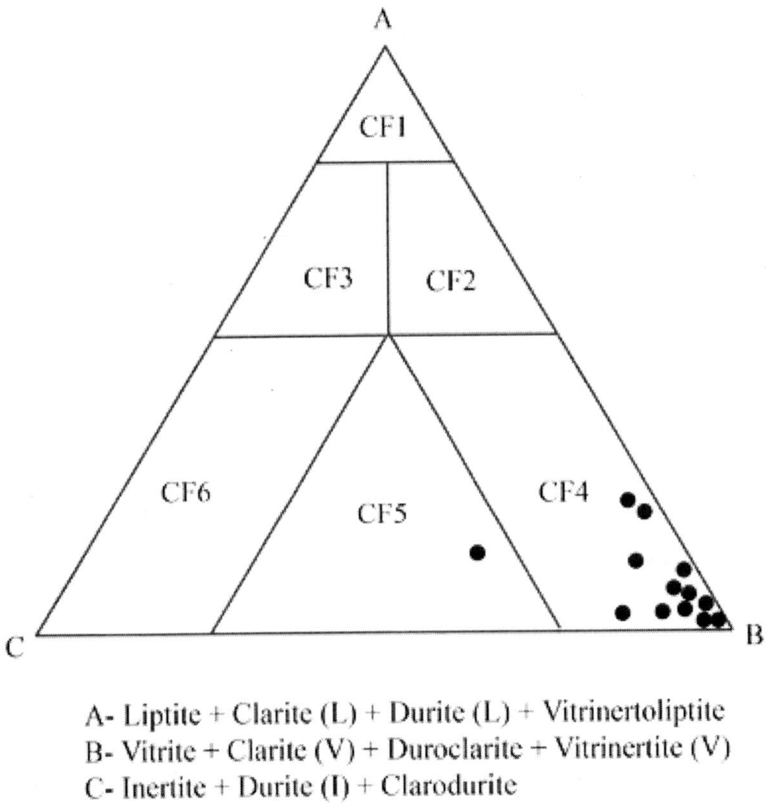


**Figure 10.** Coal facies based on microlithotype composition of Karharbari coals form the Talcher coalfields, Mahanadi valley Orissa, *CF1* deep water facies characterized by organic mud rich liptinite, *CF2* open water with dominance of sub aquatic plants, *CF3* forest swam with sudden eustatic change, *CR4* forest reed swamp, *CF5* forest swamp, *CF6* mass swamp with extremely dry conditions (Singh et al. 2013)

# 6 CONCLUSIONS

Petrographic study of the Karharbari coal shows that these coals are vitrinite rich, followed by liptinite and inertinite. Secondary liptinite macerals are observed in significant amount. The microlithotype analysis shows the dominance of vitrite followed by vitrinertite, clarite and inertite. The vitrinite reflectance (Ro) and volatile matter (d.a.f) values rank the Karharbari coal as high volatile bituminous 'C' to 'B'. The microscopic constituents of Karharbari coal indicate peat accumulation in the forest mire under fluctuating oxic to anoxic conditions with good tissue preservation.

# ACKNOWLEDGMENTS

Thanks to the Director, Rajiv Gandhi Institute of Petroleum Technology, Rae Bareli for providing all the necessary facilities. Author is also thankful to the Professor-In charge of the Coal & Organic Petrology Laboratory for extending the facilities for the coal petrography and fund from CSIR. Author is grateful to the anonymous reviewers whose constructive critiques greatly improved the manuscript.

# REFERENCES

1. Alkonde SO, Hoffknecht A, Erdtmann BD (1992) Rank and petrographic composition of selected Upper Cretaceous and Tertiary coals of Southern Nigeria. Int J Coal Geol 20:209–224
2. Amijaya H, Littke R (2005) Microfacies and depositional environment of Tertiary Tanjung Enim low rank coal, South Sumatra basin, Indonesia. Int J Coal Geol 61:197–221

3. BIS (2003) Methods of test for coal and coke (2nd revision of IS: 1350). Part I, Proximate analysis. Bureau of Indian Standard, pp. 1–29

4. Blanford WT, Blanford HF, Theobald T (1856) On the geological structure and relations of Talcher coalfield in the district of Cuttuck. Mem Geo Survey India 1:1–98

5. Calder JH, Gibbing MR, Mukhopadhay PK (1991) Peat formation in a Westphalian B piedmont setting, Cumberland Basin, Nova Scotia: Implication for the maceral-based interpretation of rheotrophic and raised paleomires. Bull De La Soc Geol De France 162:283–298

6. Cohen AD, Spackman W (1972) Methods in peat petrology and their application to reconstruction of paleoenvironments. Geol Soc Am Bull 83:129–142

7. Cohen AD, Spackman W, Raymond R (1987) Interpreting the characteristics of coal seams from chemical, physical and petrographic studies of peat deposits. In: Scott AC (ed.), Coal and coal-bearing strata: recent advances. Geological Society Special Publication No. 32, pp. 107–125

8. Das D (1959) On the fusian of Talcher coalfield, Orissa. J Geol Soc India 31:167–170

9. Das NK, Rath BD (1974) A note on the coal exploration in the western extension of Talchir coalfield. Explorer, 5–13

10. Deng XL, Sun YZ (2011) Coal petrological characteristics and coal facies of No. 11 seam from the Antaibao mine, Ningwu coalfield, China. Energy Explor Exploit 29:315–326

11. Diessel CFK (1965) On the correlation between coalfacies and depositional environment. In: Proceedings of the 20th Sydney

Basin Symposium. Department of Geology, University of Newcastle, Newcastle, pp. 19–22

12. Diessel CFK (1986) An appraisal of coal facies based on maceral characteristics. Aust Coal Geol 4:474–484

13. Diessel CFK (1992) Coal-bearing depositional systems. Springer, Berlin **721p**

14. Duan DJ, Zhao CL, Qin SJ, Kalkreuth W, Lin MY (2011) Coal petrological and coal facies characteristics of the No. 2 seam from Huangling mine, Shanxi Province, China. Energy Explor Exploit 29:647–665

15. Fox CS (1934) The lower Gondwana coalfields of India. Mem Geol Surv India 59:386p

16. Goswami S (2002) Studies on geology and palaeontology of Gondwana sequence of Ib River Coalfield, Sambalpur, Jharsuguda and Sundargarh districts, Orissa, India and their biostratigraphic significance. Ph.D. Thesis, Utkal University, Bhubaneswar, pp. 1–282

17. Gould R, Shibaoka M (1980) Some aspects of the formation and petrographic features of coal members in Australia, with special reference to the Tasman Orogenic Zone. Aust Coal Geol 2:1–29

18. Grady WC, Eble CF, Neuzil SG (1993) Brown coal maceral distributions in a modern domed tropical Indonesian peat and a comparison with maceral distributions in Middle Pennsylvanian-age appalachian bituminous coal beds. Geol Soc Am Spec Pap 286:63–82

19. Hacquebard PA, Donaldson JR (1969) Carboniferous coal deposition associated with flood plain and limnic environments in Nova Scotia. In: Dapples EC, Hopkins ME (ed), Environment of coal deposition. Geol Soc Am Spec Pap 114:143–191

20. Harvey RD, Dillon JW (1985) Maceral distributions in Illinois coals and their paleo-environmental implications. Int J Coal Geol 5:141–165

21. Hawke MI, Martini IP, Stasiuk LD (1999) A comparison of temperate and boreal peats from Ontario, Canada: possible modern analogues for Permian coals. Int J Coal Geol 41:213–238

22. Hunt JW (1989) Permian coals of eastern Australia: geological control of petrographic variation. Int J Coal Geol 12:589–634

23. Hunt JW, Smyth M (1989) Origin of inertinite rich coals of Australian cratonic basins. Int J Coal Geol 11:23–46

24. International Committee for Coal and Organic Petrology (1998) The new vitrinite classification (ICCP System 1994). Fuel 77:349–358

25. International Committee for Coal Petrology (1971) International handbook of coal petrography, 2nd edn. Centre National de recherche Scientifique, Paris

26. International Committee for Coal Petrology (2001) The new inertinite classification (ICCP System 1994). Fuel 80:459–471

27. ISO 7404-5 (1994) Methods for the petrographic analysis of bituminous coal and anthracite—Part 5: method of determining microscopically the reflectance of vitrinite. ISO, Geneva, p 11

28. Kalkreuth WD, Marchioni DL, Calder JH, Lamberson MN, Naylor RD, Paul J (1991) The relationship between coal petrography and depositional environments from selected coal basins in Canada. In: Kalkreuth WD, Bustin RM, Cameron AR (eds) Recent advances in organic petrology and geochemistry. A symposium honouring Dr P Hacquebard. Int J Coal Geol 19:21–76

29. Kizilstein LY, Minaeva LG (1972) Proiskhazhdeniji framboidlnikh form pyrite. Doki Acad Nauk USSR 206:1187–1189

30. Kizilstein LY, Trufanov VN (1968) Proiskhazhdeniji singeoeticbnogo pyrita. V ugolnikh *In:* Uglenasmniji Formatsii I Ugolnije. Mestorozhdenija. Nauka. Moskow, pp. 123–127

31. Krausel R (1961) Palaeobotanical evidence of climate. In: Nairn AEM (ed) Descriptive palaeoclimatology. Inter-Science Publishers, New York, pp 227–254

32. Lamberson MN, Bustin RM, Kalkreuth W (1991) Lithotype (maceral) composition and variation as correlated with paleo-wetland environments, Gates Formation, Karharbari British Columbia, Canada. Int J Coal Geol 18:87–124

33. Ligouis B, Doubinger J (1991) Petrology, palynology and depositional environments of the 'Grande Couche de Bourroan' from the stephanian basin of Decazeville, france. Bull De La Soc Geol De France 162:302–323

34. Lin MY, Tian L (2011) Petrographic characteristics and depositional environment of the No. 9 coal (Pennsylvanian) from the Anjialing Mine, Ningwu coalfield, China. Energy Explor Exploit 29:197–204

35. Mackowsky MT (1982) Minerals and trace elements occurring in coal. In: Stach E et al (eds) Stach's textbook of coal petrology. Gebr Borntraeger, Berlin, pp 356–361

36. Marchionni DL (1980) Petrography and depositional environments of the Liddel seam, Upper Hunter Valley, New South Wales. Int J Coal Geol 1:36–61

37. Marques M (2002) Coal facies and depositional environments of the Aurora and Cabeza de Vaca Units, Peñarroya–Belmez–Espiel Coalfield (Cordoba, Spain). Int J Coal Geol 48:197–216

38. Mishra SK, Mohanty JK, Nanda B (1998) FTIR study of lower Gondwana coals from Talcher, Orissa. J Geol Soc India 51:371–376

39. Mohanty JK, Mishra SK, Nayak BB (2001) Sequential leaching of trace elements in coal: a case study from Talcher coalfield, Orissa. J Geol Soc India 58:441–447

40. Mukhopadhyay PK (1986) Petrography of selected Wilcox and Jockson Group lignites from Tertiary of Texas. In: Finkelman RB, Casagrande DJ (ed.), Geology of gulf coast lignites. Annual Meeting in Geological Society of America, Coal Geology Division Field Trip, pp. 126–145

41. Navale GKB (1965) Petrographic analysis of some coals from Talcher coalfield. Met Miner Rev 1:5–12

42. Navale GKB (1966) Petrographical and sporological studies of some coals from Talcher coalfield Orissa, India. Paleobotanist 14:61–69

43. Navale GKB (1971) Petrological study of the coals exposed near Gopalprasad of Talcher Coalfield, Orissa, India. Paleobotanist 18:658–663

44. Neavel RC (1966) Sulfur in coal and its distribution in the seam and in mine products. Ph.D. Thesis (Unpublished). Pennsylvania State University, p. 332

45. Pareek HS (1955) Study of Talcher coals. J Sci Ind Res India 14:606

46. Pareek HS (1958) Fungal bodies in Talchir coals. J Palacont Soc India 3:214–225

47. Pareek HS (1963a) Petrology of Talcher coals. Econ Geol 58:1089–1109

48. Pareek HS (1963b) The nature and origin of certain micro-constituents in the coals of Talcher coalfield, India. C R Congr Int Strat Geol Carbonif 3:991–995

49. Pareek H S (1956) The petrology of coals from Talcher coalfield, India. Resumenes de Los Trabajos Presentadox XX Intern Geol Congr Mexico, pp. 199–200

50. Pascoe EH (1959) A manual of geology of India and Burma, 3rd edn. Government of India Press, Calcutta, p 1343

51. Raja Rao CS (1982) Bulletins of the geological survey of India, series A, No. 45, coalfields of India, coal resources of Tamil Nadu, Andhra Pradesh, Orissa and Maharastra, pp. 41–52

52. Renton JJ (1979) The mineral content of coal. In: Donaldson AC, Presley MW, Renton IJ (eds), Carboniferous coals. W.Va. Geol Surv 1:208–225

53. Reyes-Navarro J, Davis A (l976) Pyrite in coals. Its form and distribution as related to the environments of coal deposition in three selected coals From Western Pennsylvania. Special Report. No Sr.110 coal research section. Pennsylvania State University, State College, p. 141

54. Sebag D, Copard Y, Di-Giovanni Ch, Durand A, Laignel B, Ogier S, Lallier-Verges E (2006) Palynofacies as a useful tool to study the origins and transfers of particulate organic matter in recent terrestrial environments: synopsis and prospects. Earth Sci Rev 79:241–259

55. Shearer JC, Clarkson BR (1998) Whangamarino wetland: effects of lowered river levels on peat and vegetation. Int Peat J 8:52–65

56. Silva MB, Kalkreuth W, Holz M (2008) Coal petrology of coal seams from the Leão- Butiá Coalfield, Lower Permian of the

Paraná Basin, Brazil—implications for coal facies interpretations. Int J Coal Geol 73:331–358

57. Singh MP, Singh PK (1996) Petrographic characterization and evolution of the Permian coal deposits of the Rajmahal basin, Bihar, India. Int J Coal Geol 29:93–118

58. Singh MP, Singh AK (2000) Petrographic characteristics and depositional conditions of Eocene coals of platform basin, Meghalaya. Int J Coal Geol 42:315–356

59. Singh PK, Singh MP, Singh AK, Arora M (2010a) Petrographic characteristics of coal from the Lati formation, Tarakan basin, Eeast Kalimantan, Indonesia. Int J Coal Geol 81:109–116

60. Singh PK, Singh MP, Singh AK (2010b) Petro-chemical characterization and evolution of Vastan Lignite, Gujarat, India. Int J Coal Geol 82:1–16

61. Singh PK, Singh MP, Prachiti PK, Kalpana MS, Manikyamba C, Lakshminarayana G, Singh AK, Naik AS (2012a) Petrographic characteristics and carbon isotopic composition of Permian coal: Implications on depositional environment of Sattupalli coalfield, Godavari Valley, India. Int J Coal Geol 90–91:34–42

62. Singh PK, Singh MP, Singh AK, Naik AS (2012b) Petrographic and geochemical characterization of coals from Tiru valley, Nagaland, NE India. Energy Explor Exploit 30:171–192

63. Singh AK, Singh MP, Singh PK (2013) Petrological investigations of Oligocene coals from foreland basin of northeast India. Energy Explor Exploit 31:909–936

64. Skripchenko NS, Berberian TK (1975) Structura framboidalnogo pyritoc. Geo Rudn Mestarozhl 5:107–112

65. Smyth M (1979) Hydrocarbon generation in the Fly Lake, Brolga area of the Cooper basin. J Aust Petrol Assoc 19:108–114

66. Smyth M (1980) Thick coal members: products of an inflationary environment. Aust Coal Geol 2:53–72

67. Smyth M (1984) Coal microlithotypes related to sedimentary environments in the Cooper Basin, Australia. Spec Publ Int Asso Sedimentol 7:333–347

68. Styan WB, Bustin RM (1983) Petrography of some Fraser Delta peat deposits: coal maceral and microlithotype precursors in temperate-climate peats. Int J Coal Geol 2:321–370

69. Suárez-Ruiz I, Flores D, Filho JGM, Hackley PC (2012) Review and update of the applications of organic petrology: Part 1, geological applications. Int J Coal Geol 99:54–112

70. Subramanian KS (1971) Coal resources of India. Mem Geol Surv 88:575p

71. Teichmüller M (1982a) Origin of the petrographic constituents of coals. In: Stach E (ed) Stach's textbook of coal petrology. Gebrüder Borntraeger, Stuttgart, pp 219–294

72. Teichmüller M (1982b) Fluorescence microscopical changes of liptinites and vitrinites during coalification and their relationship to bitumen generation and coking behaviour. TSOP Spec Publ 1:74

73. Teichmuller M, Durand B (1983) Fluorescence microscopical rank studies on liptinites and vitrinites in peats and coals and comparison with results of the rock-eval pyrolysis. Int J Coal Geol 2:197–230

74. Teichmüller M, Teichmüller R (1975) The geological basis of coal formation. In: Stach E, Mackowsky MT, Teichmiiller M, Taylor

GH, Chandra D, Teichmiiller R (eds) Stach's textbook of coal petrology. Gebruiider Borntraeger, Stuttgart, pp 5–54

75. Teichmuller M, Teichmuller R (1982) The geological basis of coal formation. In: Stach E, Mackowsky MT, Teichmiiller M, Taylor GH, Chandra D, Teichmiiller R (eds) Stach's textbook of coal petrology. Gebruiider Borntraeger, Stuttgart, pp 5–86

76. Tiwari RS, Tripathi A, Vijaya (1995) Organicwalled microfossils of doubtful origin in Permian and Triassic sequences on Peninsular India. Palaeobotanist 43:1–38

# Chapter 10

# Geochemistry of Kasnau-Matasukh lignites, Nagaur Basin, Rajasthan (India)

**Prakash K. Singh[1] • Pramod K. Rajak[1] • Mahendra P. Singh[1] • Vijay K. Singh[1] • Amiya S. Naik[1]**

*Coal & Organic Petrology Lab, Centre of Advanced Study in Geology, Banaras Hindu University, Varanasi, Uttar Pradesh 221005, India*

## ABSTRACT

The distribution and verticals variation of geochemical components in the Kasnau-Matasukh lignites of Nagaur Basin, Rajasthan, were investigated using microscopy, proximate and ultimate analyses, Rock–Eval Pyrolysis, X-ray diffraction and Fourier Transform Infrared analyses, and major/minor/trace element determination. The relationship of elements with ash content and with macerals have also been discussed. These lignites are stratified, black, dominantly composed of huminite group macerals with subordinated amounts of liptinite and inertinite groups. They are classified as type-III kerogen and are mainly gas prone in nature. The concentration (in vol%) of mineral matter is seen to increase towards upper part of seam and so is the concentration (in wt%) of the volatile matter, elemental carbon and sulphur. The common minerals present in these lignitesare mixed clay layer, chlorite, and quartz as identified by X-ray diffraction study. Compared with world average in brown coal, the bulk concentration of Cu is anomalously high in most of the samples while Cd is 2–3 times high and Zn is high in one band. Based on interrelationship, different pyrite forms are noticed to have different preferential enrichment of various elements. The concentration of disseminated pyrite is more than the other pyrite forms and is followed by discrete pyrite grains and massive pyrite.

**Keywords :** Geochemistry Macerals Trace elements Kasnau-Matasukh lignite Rajasthan

## 1 INTRODUCTION

Coal is an organo-clastic sedimentary rock composed of lithified plant debris. The inorganic constituents occurring in different forms in coal of all rank are collectively known as 'mineral matter'. Mineral matter may occur as crystalline solids, dissolved salts in pore waters, organo-metallic compounds, discrete and disseminated grains. The type and quantity of mineral matter in coal depends on nature of vegetal matter, mode of accumulation (allochthonous or autochthonous), tectonic framework of the depositional basin, hydrological conditions, climatic conditions and the geomorphology of the hinterland. There are different phases during which the mineral matter gets into coals. These include, (i) inherent in organic matter of plants, (ii) syngenetic minerals added during the stage of peat development, and (iii) secondary or epigenetic minerals deposited through circulating waters during the coalification process. Contributions on the significance of mineral matter and trace elements in coalhave been made by many researchers. Ward et al. (1999) discussed the trace elements and mineral matter of New South Wales, Australia, in the light of quantitative data obtained through X-ray diffraction study. Mineralogical analysis of these Australian coals was further used for the seam correlation (Ward et al. 2001). A detailed account on the significance of mineral matter in coal is provided by Ward (2002). Li et al. (2010) used electron microprobe study to know the occurrence of non-mineral inorganic elements in macerals of low rank coals while Dai et al. (2013) studied the mineralogical and geochemical anomalies of Late Permian coals from southern China and evaluated the influence of hydrothermal fluids and terrigenous materials. They also carried out study on the elements and phosphorus minerals in the Jurassic coals of Tibetan Plateau (Dai et al. 2015a, b). While working on mineral matter in coal, Ren (1996) discussed the significance of trace elements in coal revealing geologic information about coal-bearing sequence formation, depositional condition and regional tectonic history of the basin. Trace elements get accumulated in coal in two ways, through plants and animals, and through geologic processes during peat and post peatification stages (Bouška et al. 2000). The impact of trace elements on environment depends upon their modes of occurrence (mobility), concentration, and toxicity (Finkelman 1995; Dai et al. 2005) and thus, study of trace elements would help in formulating strategies to combat pollution related to coal combustion (Querol et al. 2001; Dai et al. 2005; Wang et al. 2008; Tang et al. 2009). There are certain elements like As, Be, Cd, Cr, Co, Cu, Pb, Mn, Hg, Mo, Ni, Sr, U, V, and Zn which are environmentally more sensitive and have their impact on environment when released into atmosphere especially after their combustion in the thermal power plants (Pickhardt 1989; Turiel et al. 1994; Singh et al. 2011, 2012, 2014; Prachiti et al. 2011; Singh and Singh 2013). The mean abundance of elements in coal is also important for geochemical comparisons (Ketris and Yudovich 2009). Silicates, carbonates and sulphates are the major minerals in coal which consists most of the elements but some elements such as Ge, B, Br, Be, and Cl are associated with the organic matter (Finkelman 1995).

Though, various workers have contributed on the geological aspects of the Kasnau-Matasukh lignites of Nagaur basin, but no work has been carried out on the petrological and geochemical aspects of these lignites. Therefore, the present study has been undertaken to see the distribution and variation of the geochemical constituents, including major/minor/trace elements, vertically along the seam profile of these lignites. Further, the inter-relationships among the geochemical constituents and also with petrographic elements has been discussed. This would help in planning the strategy for their utilization.

## 2 GEOLOGICAL SETTING

The sedimentary tract of Rajasthan is spread over a large area of 120000 $km^2$ and forms the eastern flank of Indus shelf. The entire sedimentary tract has been sub-divided into four basins which includes- (i) Palana-Nagaur basin, (ii) Jaisalmer basin, (iii) Barmer basin, and (iv) Sanchor basin (Jodha 2008). The present investigation has been carried out on Kasnau-Matasukh lignites. These lignites occur in Palana-Nagaur basin which is an E-W trending elongated basin. It extends for 200 km in length and 50 km in width. Kasnau-Matasukh block is located in Jayal Tehsil of Nagaur district. Nagaur basin has several disconnected small basins of Palana Formation indicating undulating paleotopography. Nagaur basin is linked to a 5–6 m wide channel and hence known as link basin (Lal and Regar 1991). Though the structural features indicate a low tectonic disturbance in the area, gravity survey has indicated the presence of gravity low of 3–4 km width which occurs in NW–SE direction indicating the presence of link channel broadening towards SE. The magnetic survey of the area also substantiates the subsurface structures delineated through gravity survey and show NW–SE elongation of magnetic contour (Lal and Regar 1991).Geological work in and around Nagaur was initially taken up by Blanford (1876) who correlated the Jodhpur set with Vindhyans because of their closer resemblance. The interpretation of exploration data of Nagaur basin has been incorporated in the reports of Mukhopadhyay (1974–1975), Munshi (1975–1977), Faruqi (1978–1979, 1982–1983).

The lignite bearing Lower Tertiary sediments of Palana Formation are deposited unconformably over the Nagaur Formation. Presence of one lignite seam has been identified in the block which is intersected by a number of dirt bands. The Tertiary sequence of rocks comprises of three Formations which include Palana, Marh and Jogira in ascending order (Jodha 2009). The area has scanty outcrops. The lignite occurrences in the Nagaur basin are associated with Palana Formation of Paleocene age. The seams have been reported between 50 m and 150 m depths in the Palana Formation (Jodha 2009). Based on the recovered palynomorphs comprising pteridophytes, angiosperms, algae and fungi, Kulshrestha et al. (1989) and Shah and Kar (1971) have given Paleocene age to this Formation. The general stratigraphic succession of the rocks in the basin is given in Table 1 and the general geological map is shown in Fig. 1. The litholog (after Ghose 1983) and megascopic profile of the Kasnau-Matasukh lignite seam, prepared for this study, is shown in Fig. 2a while geological section is shown in Fig. 2b.

**Table 1.** General stratigraphic succession in the Bikaner-Nagaur basin, Rajasthan (after Ghose, 1983)

| Age | Formation | Lithounits | Thickness (m) |
|---|---|---|---|
| Pleistocene to recent | Kolayat formation | Sand and sandy alluvium | 5–11 |
| | | Ironstone nodule, sandy calcareous grit kankar, gypsite, Ferruginous band, semi-consolidated conglomerate | 1–2 |
| | | Erratic boulder of quartzite | ? |
| Unconformity | | | |
| Early to middle eocene | Jogaria formation (Calcareous facies) | Shaly and marly limestone with foraminifers (*Alveolina, Discocyclina Nummulities*) | 5–10 |
| | | Unfossiliferous, white clayey marl | 1 |
| | | Dirty brown impure limestone with broken shells of *ostrea* and foraminifers (*Assilina*) | 1.5 |
| | | Fuller's earth with shale partings having casts of lamellibranchs and gastropods | 14 |
| | | Cream and yellowish white limestone full of smaller foraminifers (*Nummulites*and *Assilina*) with a thin band of fuller's earth (1–2 m) near base | 75 |
| | | Yellow shales ochers, marl, etc. with smaller foraminifers (*Nummulites,Assilina*) | 20 |
| Angular unconformity | | | |
| Late paleocene (?) | Marh formation (Arenaceous facies) | Upper clay horizon with one clay bed | 3–10 |
| | | Ferruginous sandstone, gritty sandstone and sugary sandstone with white glass sand (local) | 60 |
| | | Middle clay horizon with five clay beds and sandstone partings | 50 |
| | | Ferruginous sandstone, gritty sandstone, grit, siltstone | 70 |
| | | Lower clay horizon with one clay bed | 1–3 |
| | | Ferruginous sandstone, gritty sandstone, various siltstone with leaf impressions (base not exposed) | 20 |
| (?)Gradational contact | | | |
| Early paleocene (?) | Palana formation (Carbonaceous facies) | Fine grained sandstone Carbonaceous shale and lignite | ? |
| Base not encountered | | | |

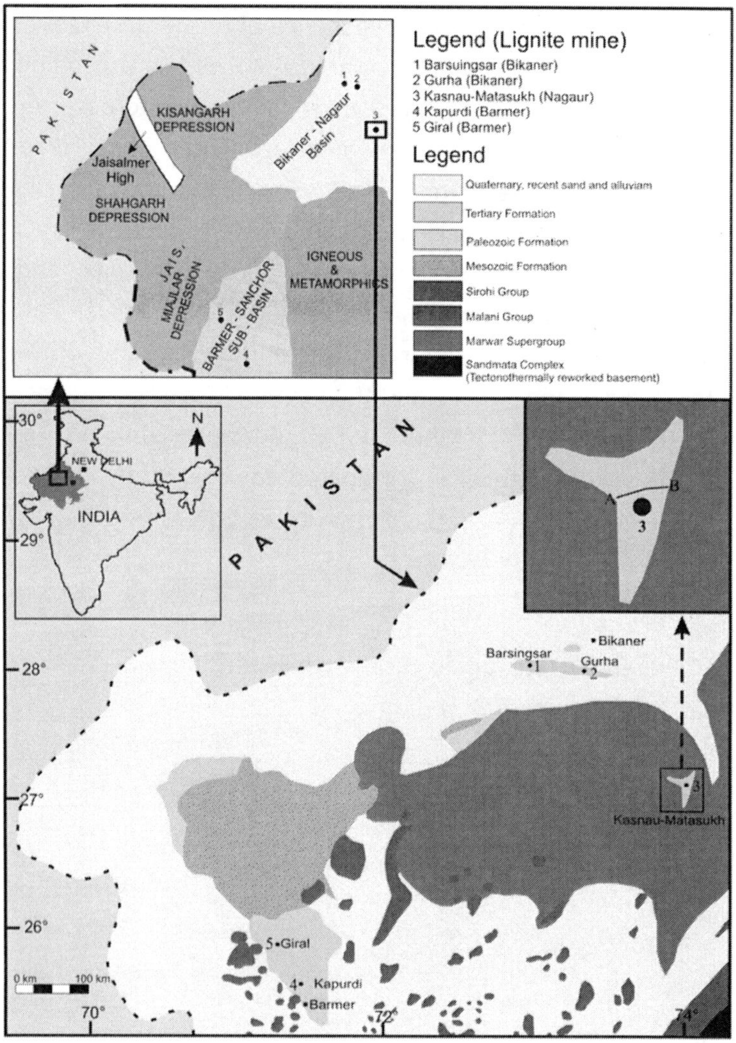


**Figure 1.** Geological map of Nagaur basin (after Roy and Jakhar 2002) and location of Kasnau-Matasukh lignite

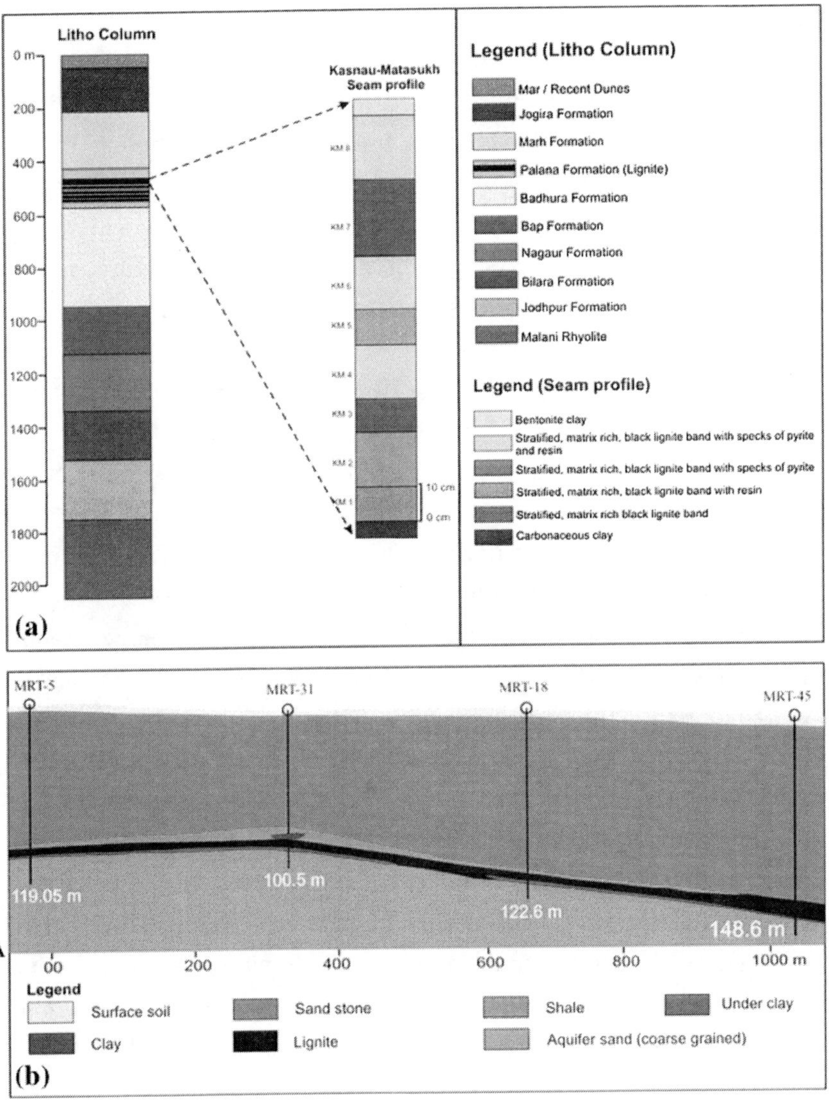


**Figure 2.** Litholog (after Ghose 1983), megascopic profile of the Kasnau-Matasukh lignite seam showing various bands (**a**) and geological section (**b**) which is redrawn after Gowrisankaran et al. (1987)

## 3 METHOD OF STUDY

Lignite samples have been collected from Kasnau-Matasukh mine of Nagaur basin of Rajasthan (Fig. 1) following pillar sampling method (Schopf 1960) so that full lignite seam thickness may be reconstructed in the laboratory. The samples have been crushed and reduced in quantity through quartering and coning to prepare eight composite samples which were subjected to various analyses. The samples were ground to pass 18 mesh size to prepare polished mounts for petrography, while they were further ground to pass 70 mesh size for various chemical analyses like proximate, ultimate, rock–eval pyrolysis and major/minor/trace element analyses. Maceral analysis has been carried out to see the distribution of huminite, liptinite and inertinite group macerals. This is performed under reflected light using a LeitzOrthoplan-Pol Microscope equipped with Wild Photoautomat MPS 45 in the Coal and Organic Petrology Laboratory, Department of Geology, Banaras Hindu University. The line-to-line and point-to-point spacing was maintained at 0.4 mm and more than 600 counts have been taken on each sample following the methodology given by Taylor et al. (1998); huminite macerals have been termed and described as per ICCP-1994 (Sýkorová et al. 2005) while ICCP (2001) has been followed for inertinite macerals. The vitrinite/huminite reflectance (VRo) was measured at National Metallurgical Laboratory, Jamshedpur following ISO 7404-5:2009 (standard used: spinel, yag-yittrium, aluminium garnet, zirconia). On each sample, a minimum of 200 measurements were taken. The proximate analysis has been carried out as per BIS (2003), while the elemental analysis (C, H, N, O, and S) has been performed at CMPDI, Ranchion Elementar Analysensysteme-Vario-III as per ASTM D5373-08.

The Pyrolysis has been carried out on high precision Rock–Eval-6 (make Vinci Technologies, France) at R & D department, Oil India Ltd, Duliajan (Assam) on fourteen lignite samples (represented as four composite samples).This is programmed pyrolysis system and is ultimate to know the source rock potential for hydrocarbon. The significance of this technique is that the coal samples, as such,are analyzed to know the various components. The analysis is performed under controlled temperature and coal samples are heated in absence of oxygen. The produced compounds are quantitatively assessed. During heating the oxygenated compounds released from mineral matter, present in coal, are excluded. The quantitative measurement of various fractions of volatile/non volatile organic compounds, source rock potential and the degree of maturation of lignites samples is obtained through this analysis. The pyrolysis of Kasnau-Matasukh lignite samples has been carried out following the procedures of Espitalié et al. (1977, 1984, 1986). The samples were heated in an open pyrolysis system under non-isothermal condition and the recorded FID signal is divided in two surfaces, $S_1$ and $S_2$, which are expressed in mg HC/g of coal. The method gets completed by combustion (oxidation) of the residual rock recovered after pyrolysis at 850 °C under nitrogen. This is required to avoid incomplete combustion. The released CO and $CO_2$ are monitored online through an infra-red cell. This complementary data acquisition helps in the determination of total organic carbon (TOC) and total mineral or inorganic carbon (TMC or TIC).

The elements Fe, Ca, Mg, Mn, K, Na, Cu, Co, Ni, Cr, Zn, Pb and Ashave been determined on 'whole coal samples' in the department of Botany, BHU, Varanasi. For the determination of these elements the coal samples have been digested with 2.5 mL $HNO_3$ and$HClO_4$ in 10:1 ratio on hot water plate following the method ofEaton et al. (1995).The mixture is then filtered usingWhatmanfilter paper (No. 41) and the digested samples are rinsed with 1 % Conc. $HNO_3$.It is then transferred in a separate test tube and the volume is made up to 20 mL. The digested samples have been used for analyzing the concentrations of various elements under Atomic Absorption Spectrophotometer (AAS, Model Perkin Elmer Analyst 800) and the standard used in the analysis were Accu Standard solutions obtained from Merck, KGaA, Darmstadt, Germany. Data of major, minor and trace elements are mean of three independent observations in the present paper. The measured values have shown relative standard deviations less than 5 % for all the elements in the analyzed samples.

Fourier Transform Infrared spectra have been recorded by FTIR spectrophotometer (PerkinElmer Spectrum version 10.03.05) using KBr pellets (transmission mode) in the Department of Chemistry, Banaras Hindu University. Coal: KBr mixture at 1:100 ratio has been used and 20 number of scans have been taken with a spectral resolution of 4 $cm^{-1}$ at a range of 400–4000 $cm^{-1}$. X-ray diffraction data have been obtained with the help of computer controlled Xray Diffractometer PanalyticalX'Pert High Score (Plus) v3× database in the Department of Geology, Banaras Hindu University. The operating parameters, in the present study are: start angle- 2°; target- Cu Kα radiation; stop angle-60°; step size- 0.0250; and 2 theta configuration.

# 4 RESULT AND DISCUSSION

## 4.1 Petrographic characteristics

These lignites are stratified in nature and are of black color. Huminite is the main component in these lignites (Singh et al. 2015a, b) which is formed due to anaerobic preservation of lignocellulose material in the mire (Sýkorová et al. 2005). Liptinite and inertinitemacerals occur in low concentrations. Huminite (83.9 %–92.5 %; av. 87.3 % mineral matter free basis)is largely contributed by detrohuminite and telohuminite. Detrohuminiteis represented by densinite (19.2 %–42.5 %; av. 31.7 % mineral matter free basis) and attrinite (0 %–13.3 %; av. 5.9 % mineral matter free basis) while telohuminiteis represented by ulminite-A (24.9 %–38.6 %; av. 30.5 % mineral matter free basis),ulminite-B (13.4 %–29.1 %; av. 18.2 % mineral matter free basis) and textinite which occurs in very low amount (<1 %). Liptinite group (5.7 %–13.2 %; av. 10.9 %) and inertinite group (0.2 %–4.0 %; av. 1.9 %) are low in concentrations (Table 2). Mineral matter ranges between 3.5 and 12.0 (av. 7.7 %). The vertical variation of group macerals and mineral matter from base of the seam is shown in Fig. 3. Though, there is no specific trend of variation, yet huminite shows a high concentration at the upper part while liptinite shows a reverse trend. Inertinite is less at the bottom.

Mineral matter is more towards the upper part of the seam. The variation has environmental implications. The clastic mineral matter relates directly to water cover in the basin and, therefore, it increases with increase in the water cover during the formation of upper part of the lignite seam. This is also supported by the occurrence of high concentration of huminite group macerals during this period.

## 4.2 Chemical attributes

Theseligniteshave high volatile matter content (52.6 %–67.0 % daf basis; av. 58.3 %) with moderate ash yield (3.0 %–18.2 %; av. 8.4 %). The ultimate analysis (av. values on daf basis) shows that theselignites contain 54.0 % carbon, 5.4 % hydrogen, 0.8 % nitrogen, 35.9 % oxygen and 3.8 % sulfur (Table 2). The vertical variation of the chemical components along the seam profile is shown in Fig. 4. Volatile matter shows an increasing trend towards upper part of seam while fixed carbon shows a reverse trend. Carbon and sulfur show an increasing trend towards upper part while other ultimate components like hydrogen, nitrogen and oxygen do not show any definite trend. Variable concentrations and dimensions of undecomposed, partly decomposed

**Table 2.** Petrographic and chemical constituents in the Kasnau-Matasukh lignites

| Sample number | Megascopic characteristics | Petrographic components (vol%) | | | | Proximate components (wt%) | | | Ultimate components (wt%) | | | | | H/C | O/C |
|---|---|---|---|---|---|---|---|---|---|---|---|---|---|---|---|
| | | Huminite | Liptinite | Inertinite | Mineral matter | Air dried | Dried ash free | | Dried ash free | | | | | | |
| | | | | | | Ash | Volatile matter | Fixed carbon | C | H | N | O | S | | |
| (Top) KM 8 | Stratified, matrix rich, black inhomogeneous lignite with presence of pyrite and resin | 82.7 (92.5) | 5.1 (5.7) | 1.6 (1.8) | 10.6 | 11.5 | 67.0 | 33.0 | 52.6 | 5.6 | 0.8 | 36.5 | 4.4 | 1.27 | 0.52 |
| KM 7 | Stratified, matrix rich, black inhomogeneous lignite with presence of pyrite | 78.2 (84.8) | 12.1 (13.1) | 2.0 (2.1) | 7.8 | 7.5 | 54.6 | 45.4 | | | | | | | |
| KM 6 | Stratified, matrix rich, black inhomogeneous lignite with presence of pyrite and resin | 76.2 (86.6) | 10.8 (12.3) | 1.0 (1.1) | 12.0 | 18.2 | 58.4 | 41.6 | | | | | | | |
| KM 5 | Stratified, matrix rich, black inhomogeneous lignite with presence of resin | 82.5 (87.0) | 10.8 (11.4) | 1.5 (1.6) | 5.2 | 4.9 | 63.3 | 36.7 | 53.7 | 5.1 | 0.8 | 36.5 | 3.8 | 1.14 | 0.51 |
| KM 4 | Stratified, matrix rich, black inhomogeneous lignite with presence of pyrite and resin | 80.9 (89.3) | 8.5 (9.4) | 1.2 (1.3) | 9.5 | 9.6 | 57.5 | 42.5 | | | | | | | |
| KM 3 | Stratified, matrix rich, black inhomogeneous lignite with presence of pyrite | 80.6 (86.9) | 8.4 (9.0) | 3.7 (4.0) | 7.2 | 7.1 | 58.2 | 41.8 | | | | | | | |
| KM 2 | Stratified, matrix rich, black inhomogeneous lignite | 80.9 (83.9) | 12.8 (13.2) | 2.8 (2.8) | 3.5 | 3.0 | 52.6 | 47.4 | | | | | | | |
| (Bottom) KM 1 | Stratified, matrix rich, black inhomogeneous lignite | 82.5 (87.2) | 11.9 (12.6) | 0.2 (0.2) | 5.4 | 5.1 | 54.3 | 45.7 | 55.7 | 5.6 | 0.9 | 34.6 | 3.2 | 1.19 | 0.47 |
| Mean | | 80.6 (87.3) | 10.1 (10.8) | 1.7 (1.9) | 7.7 | 8.4 | 58.3 | 41.8 | 54.0 | 5.4 | 0.8 | 35.9 | 3.8 | 1.20 | 0.50 |

*Note* Values in parenthesis are recalculated on m.m.f. basis

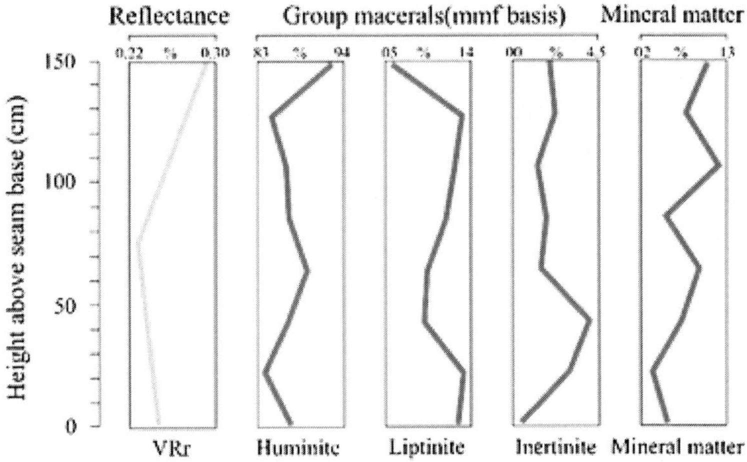


**Figure 3.** Vertical variation of group macerals and mineral matter along lignite seam profile

and completely decomposed wood have been noticed in the Kasnau-Matasukh lignites. This appears to have affected the proximate and ultimate composition of this lignite from bottom to top because these three components have variation in the organic geochemical constitution.

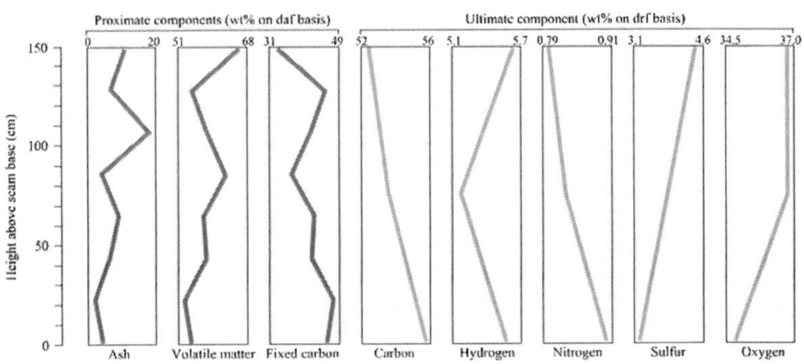


**Figure 4.** Vertical variation of geochemical components from base of the lignite seam

### 4.3 Hydrocarbon potential

These lignites of Nagaur basin have attained a thermal maturity indicated by vitrinite reflectance (VRo) between 0.23 % and 0.30 % (Table 3) which put them as 'low rank C' coals as per ISO-11760 (2005). The analytical results of Rock–Eval pyrolysis of Kasnau-Matasukh lignites show that $S_1$ values (free hydrocarbon distilled out of samples at initial heating of 300 °C) vary from 1.16 to 2.83 mg HC/g. Considering 1 mg HC/g as its cut-off value,t his lignite may be considered as a good source rock. Similarly, $S_2$ values (hydrocarbons generated through thermal cracking which actually indicate the quantity of hydrocarbons that the lignite may potentially produce) are many-fold higher than the free hydrocarbons (already generated oil in the lignite and occur as free hydrocarbons in lignite samples) and it varies from 48.65 to 87.84 mg HC/g (av. 68.99 mg HC/g). Taking 5 as its cut-off value, it also indicates a good source rock for hydrocarbon generation. The $S_3$ values vary from 24.34 to 29.89 mg $CO_2$/g and represent the trapped carbon-di-oxide which is released during the pyrolysis up to a temperature of 390 °C. This is also proportional with the oxygen present in the Kasnau-Matasukh lignites of Nagaur basin. Total organic carbon (TOC) content of these lignite samples exhibits a wide range from 3.27 % to 43.92 % with an average of 31.08 % while the total inorganic carbon (TIC) values also have the similar trend and the value ranges from 1.85 % to 2.73 %.

**Table 3.** Vitrinite/huminite reflectance data in the Kasnau-Matasukh lignites

| Sample number | Mean value | SD | Variance | Maximum | Minimum |
|---|---|---|---|---|---|
| KM-1 | 0.25 | 0.06 | 0.00 | 0.41 | 0.09 |
| KM-5 | 0.23 | 0.06 | 0.00 | 0.40 | 0.09 |
| KM-8 | 0.30 | 0.07 | 0.00 | 0.46 | 0.15 |
| Mean (Mine) | 0.26 | 0.06 | 0.00 | 0.43 | 0.11 |

The vertical variation of the Rock-Evaldata from the base of the lignite seam is shown in Fig. 5. It is evident from this figure and the data (Table 4), that $S_1$ valueis low at the bottom while $S_2$ value is more at the middle part of the seamwhich decreases towards the top as well as towards the bottom part (Table 4; Fig. 5). Total organic carbon content has an increasing trend towards the mid of the seam while decreases at the upper part. The total inorganic carbon (TIC) values also have the similar trend and are derived from the carbonates. In this seam the sulfur content (varies from 3.2 % to 4.4 %) maintains a strong negative correlation ($r = -0.81$; P value = 0.399) with the TOC content and also with TIC ($r = -0.78$; P value = 0.433). The organic matter (OM, obtained by deducting ash content from hundred) shows a variation from 81.8 % to 97 %.

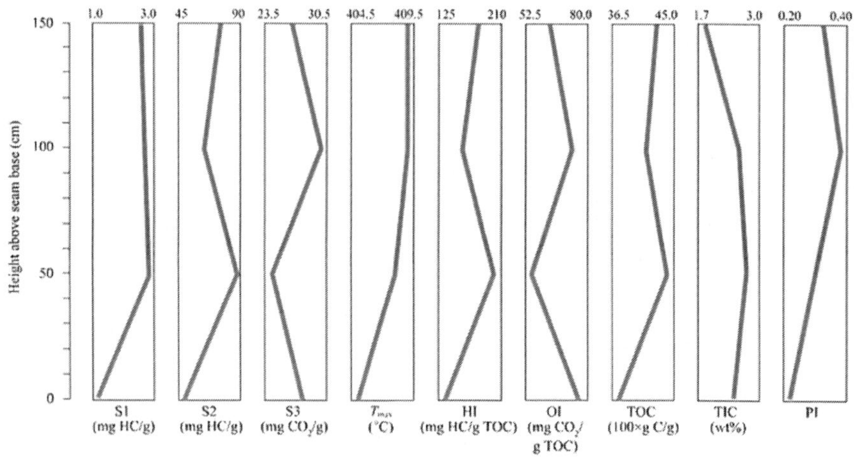


**Figure 5.** Vertical variation of Rock–Eval components from base of the lignite seam

**Table 4.** Rock–Eval data in the Kasnau-Matasukh lignites

| Sample number | $S_1$ (mg HC/g) | $S_2$ (mg HC/g) | $S_3$ (mg $CO_2$/g) | $T_{max}$(°C) | HI | OI | TOC(%) | TIC | PI |
|---|---|---|---|---|---|---|---|---|---|
| KM-8 | 2.58 | 75.76 | 26.59 | 409 | 179 | 63 | 42.3 | 1.85 | 0.03 |
| KM-5 | 2.69 | 63.71 | 29.89 | 409 | 157 | 73 | 40.67 | 2.57 | 0.04 |
| KM-3 | 2.83 | 87.84 | 24.34 | 408 | 200 | 55 | 43.92 | 2.73 | 0.03 |
| KM-1 | 1.16 | 48.65 | 27.81 | 405 | 133 | 76 | 36.46 | 2.45 | 0.02 |
| Mean | 2.32 | 68.99 | 27.16 | 407.75 | 167.25 | 66.75 | 40.84 | 2.40 | 0.03 |

*Note HI* hydrogen index, *OI* oxygen index, *TOC* total organic carbon, *TIC* total inorganic carbon

Coal acts as good source rock for hydrocarbon generation. The H/C ratio, in 0.8–0.9 range, is a good indicator of a source rock having hydrocarbon potential (Powell and Boreham 1994). Certain coals with low liptinite content have hydrogen-rich vitrinite which generates oil (Bertrand 1989; Newman et al. 1997; Petersen et al. 2000; Singh 2012; Singh et al. 2013). The generated hydrocarbon products have a finite storage capacity and until this capacity is exceeded, no oil expulsion takes place (Powell 1978; Mc Auliffe 1979; Durand 1983; Tissot and Welte 1984; Inan et al. 1998). Singh (2012) and Singh et al. (2016) have studied the hydrocarbon potential of lignites of Cambay basin and Bikaner basin (India) respectively while Singh et al. (2013) investigated the sub-bituminous coals of east Kalimantan (Indonesia) for its liquid hydrocarbon potential. The cross plot of hydrogen index (HI) with oxygen index (OI) and $T_{max}$ of Kasnau-Matasukh lignite (Fig. 6) indicates its immaturity. This lignite falls closer to the zone of organic rich type-III kerogen which is formed under topogenous condition as also revealed by a cross plot between total organic carbon (TOC) and sulfur content (Fig. 7). This plot is proposed by Jasper et al. (2010) who have grouped the coals in three domains. The coal, in Group-A, is for high TOC and low sulphur content (<2 %) which evolves under ombrogenous mires having raised bogs. Such mires are fed by rainwater. Group-C is characterized by coals with high sulphur content and is formed under topogenous mires. Group-B, however, shows an intermediate condition. On this plot the Kasnau-Matasukh lignite falls under Group-C which is characterized by coals having high sulphur content and is formed under topogenous mires as per Jasper et al. (2010). The H/C atomic ratio of 1.2 (Table 2),and high concentration (>80 %) of reactive macerals (huminite + liptinite), in Kasnau-Matasukh lignites (Table 2), are indication of their good hydrocarbon potential as per Cudmore (1977) and Davis et al. (1976). These ligniteshave potential of generating mainly gaseous hydrocarbons. The cross plot between vitrinite reflectance and HI (Fig. 8) also indicates that these lignites are mainly gas prone. The details of the maturity and oil generating potential of the lignites of entire Bikaner-Nagaur basin have been discussed in detail by Singh et al. (2016).

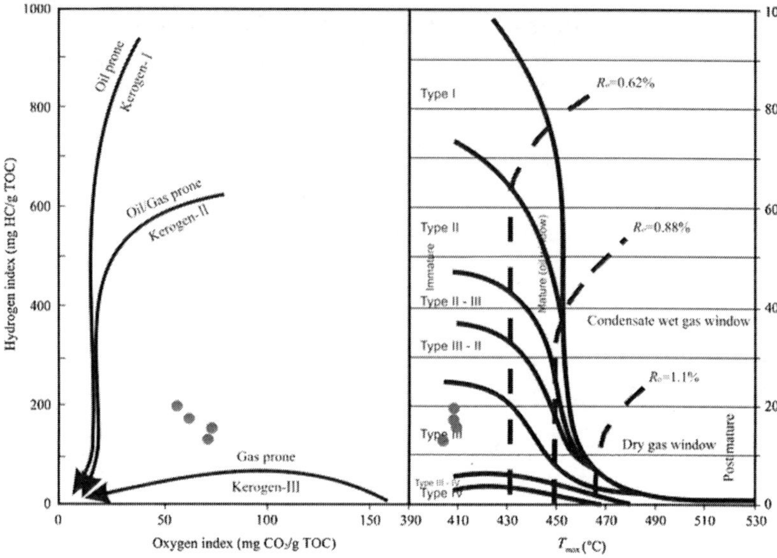


**Figure 6.** Cross plot of hydrogen index (HI) with oxygen index (OI) (Krevelen 1961) and $T_{max}$ (°C) (after Koeverdon et al. 2011)

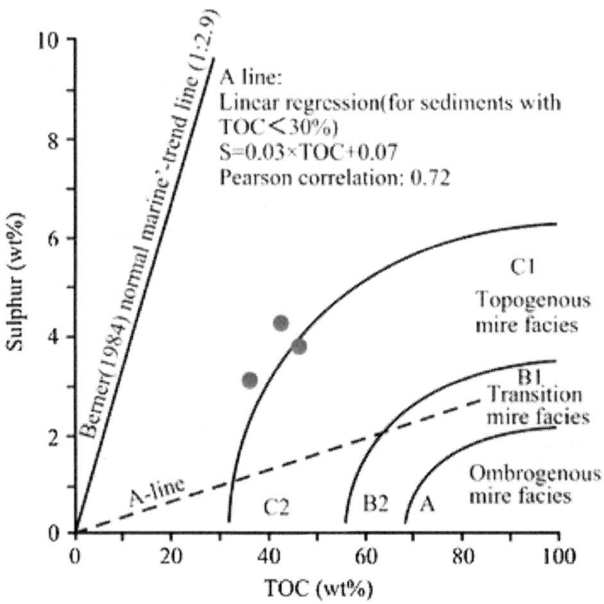


**Figure 7.** Cross plot between Sulfur and total organic carbon (TOC) (after Jasper et al. 2010)

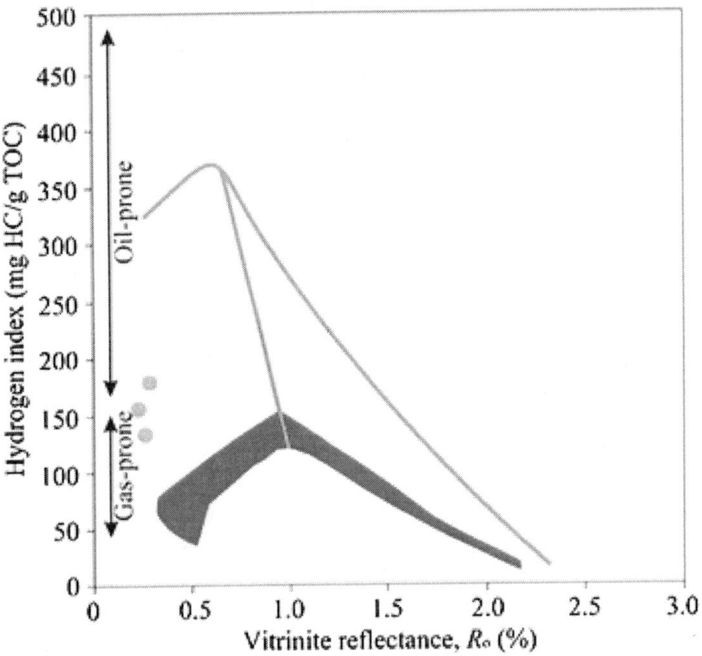


**Figure 8.** Cross plot between Hydrogen Index (HI) and Vitrinite Reflectance (VRo) (after Petersen 2005)

## 4.4 X-ray diffraction (XRD) and Fourier transform infrared (FTIR) studies

XRD spectra of whole coal and low temperature ash samples of the Kasnau-Matasukh lignites are shown in Fig. 9a, b. The minerals in these coals were identified by comparing 'd' values as per Lindholm (1987). The common minerals identified from XRD spectrum of whole coal sample are biotite, gypsum, chlorite, goethite/laumontite, quartz, barite, dolomite, haematite and marcasite. The minerals identified in the low temperature ash include goethite/laumontite, anorthite, quartz, haematite and mixed clay. Kaolinite, illite and chlorite are the major mixed clay minerals. Haematite, goethite and marcasite are major iron containing minerals while gypsum, dolomite and laumontite are calcium rich. FTIR spectra are useful for the identification of minerals associated

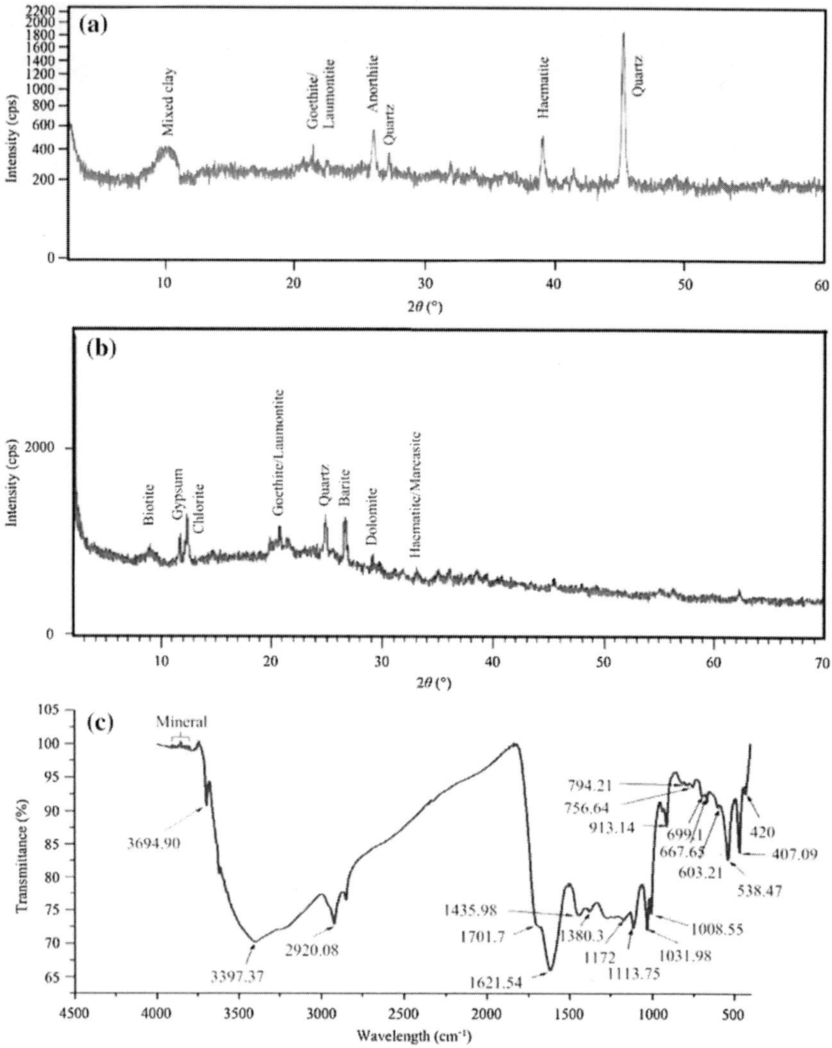


**Figure 9.** X-ray diffraction pattern and FTIR spectrum of the lignite samples of Kasnau-Matasukh

with the coal structures (Karr 1978). The peaks in FTIR spectra of coal between 1100 and 400 $cm^{-1}$ are of clay minerals such as quartz, kaolinite, illite and montmorillonite groups. The different absorption peaks with their bonds and functional groups are furnished in Table 5 and shown in Fig. 9c. The broad absorption bands in coal ranging from 3618 to 3628 and 3696 to 3699 $cm^{-1}$ belong to clay minerals (kaolinite

**Table 5.** Band assignment of the most prominent peaks in the FTIR spectra of Kasnau-Matasukh lignite

| Bands ($cm^{-1}$) | Assignments |
|---|---|
| 466–470 and 528–535 | Clay and silicate minerals |
| 669 | Aromatic out of plane bebding (?) |
| 690–515 | Alkyl halides (C–Br stretch) |
| 700–610 | Alkynes (–C≡C–H: C–H bend) |
| 725–720 | Alkanes (C–H rock) |
| 850–550 | Alkyl halides (C–Cl stretch) |
| 900–675 | Aromatics (C–H "oop") |
| 910–665 | 1°, 2° amines (N–H wag) |
| 950–910 | Carboxylic acids (O–H bend) |
| 1032–1047 and 1115–1118 | Silicate (Si–O), C–O and C–O–R structures |
| 1250–1020 | Aliphatic amines (C–N stretch) |
| 1300–1150 | Alkyl halides C–H wag (–CH2X) |
| 1370–1350 | C–H rock (alkanes) |
| 1500–1400 | Aromatics C–C stretch (in–ring) |
| 1612–1622 | C=C aromatic stretching and C=O conjugated |
| 1650–1580 | 1° amines (N–H bend) |
| 1710–1665 | $\alpha,\beta$-unsaturated aldehydes, ketones (C=O stretch) |
| 2851 | Aliphatic symmetric –CH2 stretching |
| 2918–2926 | Aliphatic asymmetric –CH2 stretching |
| 3000–2850 | Alkanes (C–H stretch) |
| 3402–3416 | –OH stretching |
| 3400–3250 | 1°, 2° amines, amides (N–H stretch) |
| 3640–3610 | Alcohols and phenols (O–H stretch, free hydroxyl) |
| 3618–3628 and 3696–3699 | Clay minerals (kaolinite and illite) |

and illite).The absorption bands at 3694.90 and 2920.08 $cm^{-1}$ in the coals are due to O–H groups while the absorption bands at 3397.37 $cm^{-1}$ are due to O–H and N–H groups. Strong aliphatic absorptions are observed at 2920–2850 $cm^{-1}$. The intensity of peaks at 2920 $cm^{-1}$ indicates the presence of long aliphatic chains in the Kasnau-Matasukh lignites. Low intensity aromatic bands were observed in 900–700 $cm^{-1}$ regions in these lignites. The peak 1701.7 $cm^{-1}$ appears due to the presence of carbonyl (C=O) group while the peak 1621.54 $cm^{-1}$ appears due to the presence of 1° amines (N–H). The peak 1435.98 $cm^{-1}$

indicates the presence of aromatics (C–O stretch in ring). The peaks near 1172 and 1113.75 $cm^{-1}$ indicate the presence of aliphatic amines (C–N). The oxygen containing functional groups are phenols, alcohols, ethers, carboxylic acid and carbonyls. The region of 1000–1300 $cm^{-1}$ in the spectra is of C–O bonds. The weak band at 756.64 $cm^{-1}$ could be due to C–Cl bond while the weak band at 699.1 $cm^{-1}$ may be due to C≡C–H; C–H bonds. The peak at 667.65 $cm^{-1}$ may be due to C≡C–H; C–H bonds. The weak bands at 603.21 and 538.47 $cm^{-1}$ is due to C–Br bond. The present study is in agreement with the studies of Georgakopoulos et al. (2003), Saikia et al. (2007) and Zodrow et al. (2010).

## 4.5 Geochemistry of major/minor and trace elements

The study of trace elements in coal is being given more impetus during last few decades owing to their environmental implications. Mode of occurrence of major, minor and trace elements in coal may be known through direct and indirect methods (Eskenzy and Stefanova 2007). In the lignite samples of Kasnau-Matasukh, the mode of occurrence of elements has been studied through indirect method. Here, correlation coefficients of the elements with ash yield, petrographic content and also among themselves have been calculated.

The concentration of elements in the analysed samples has been compared with the world average in lignite. As we can see from the Table 6 the concentration of Cu is very high in all the bands and over 70 times, in KM-7 band, as compared to world average in brown coals. Similarly, Cd is 2–3 times high in almost all the bands while Zn is high in KM-3 band. Rests of the elements have a normal concentration in Kasnau-Matasukh lignites. The vertical variation of various major/minor and trace elements is shown in Fig. 10 along the lignite seam profile. Though, there is not a prominent trend of distribution of these elements yet the concentration of elements like Mn, Na, Cu, Ni, Co, Cr, Pb and Cd is higher towards the upper part of the seam as revealed in Fig. 10. Sulfur concentration is high in Kasnau-Matasukh lignites. Pyrite is formed, in coal, from $H_2S$ and Fe in solution which involves bacterial reduction of $SO_4$ to $H_2S$ at pH 7–4.5 (Ryan and Ledda 1997). It occurs in various forms in Kasnau-Matasukh lignites. As analyzed under microscope, disseminated pyrite in these lignites, dominates (av. 41 %) over the other pyrite forms and is followed by discrete pyrite grains (av. 23.8 %) and massive pyrite (av. 11.7 %) (Table 7). Some photomicrographs of pyrite of Kasnau-Matasukh lignites are shown in Fig. 11. The clustered framboidal pyrites are more common in the middle part of the seam while single framboids are more towards the upper part. Based on the values of correlation coefficient, preferential enrichment of Ni, Pb, and Co is seen in pyrite. Finkelman (1994) also has reported the association of Co with sulfides, though, it is also found associated with clays and organic matter. Co may also occur as siegenite and cattierite (Dai et al. 2015a, b).Dale et al. (1999) have reported the occurrence of Co associated with silicates in Australian coals. Framboidal pyrite has shown preferential enrichment of Cu, Pb, Co, Cr, and Ni; disseminated pyrite shows an affinity with Ni and Co while discrete pyrite grains with Pb and Co. Similarly massive pyrite has a close affinity with Fe and Zn while pyrite occurring as fissure and crack fillings has affinity with Cd and Mg. Cadmium is normally associated with sphalerite (ZnS) though it is also found in other sulphides (Finkelman 1994). This has also been documented by Swaine (1990), Goodarzi (2002), and Dale et al. (1999). Ash yield shows a strong affinity with Mn ($r = 0.728$) and Na ($r = 0.744$) among the major elements and with Pb ($r = 0.786$) and Co ($r = 0.65$) among the trace elements. Eskenzy (2009) also reported a positive correlation of ash content with Mn and Co and observed the association of Pb with organic as well as inorganic fractions in Bulgarian coal. On the other hand inertinite maceral group has shown a strong affinity with Mg and Zn while huminite has a strong affinity with Mn, and liptinite relates well with Cu and Cr. These elements could either be associated with the organic molecules or with the minerals occurring as intergrown with the macerals. On the other hand some elements could be related to those minerals which occur as surface blanketing or as superficial mounting over the surface of the macerals (Singh et al. 2010). While working on Shenbei Tertiary lignites of China, Ren et al. (2004) reported Cr, Co, Ni, Cu, V and Zn to be associated with organic macromolecules and they have suggested their enrichment during coal-forming or early diagenesis process. Eskenzy and Stefanova (2007) believe that organically bound parts of the elements are generally higher in low-rank coals. Due to lack of evidence regarding mode of occurrence of Ni in coal, its relation is yet to be precisely established (Finkelman 1994; Riley et al. 2012). It may be organically bound or it could also be associated with sulfides. Dale et al. (1999) reported Ni from both monosulphides and organic matter. Ni relates with Na, K and Co in Kasnau-Matasukh lignites which is in agreement with the work of Singh et al. (2015a, b) on the nearly located Barsingsar and Gurhalignites of Bikaner-Nagaur basin. KM-3, which is matrix rich stratified band, contains high concentration of Zn, Cu and Cd. Zinc is considered as a notorious contaminant and occurs in all coals in HCl soluble phase (Riley et al. 2012).

**Table 6.** Major, minor and trace element contents (in ppm) in the Kasnau-Matasukh lignites

| Metal | WCBC* | Sample number | | | | | | | |
|---|---|---|---|---|---|---|---|---|---|
| | | KM-1 | KM-2 | KM-3 | KM-4 | KM-5 | KM-6 | KM-7 | KM-8 |
| Fe | 10000 | 388.00 | 414.00 | 392.00 | 61.20 | 432.00 | 418.00 | 472.00 | 506.00 |
| Cu | 15 | 342.00 | 344.00 | 374.00 | 232.00 | 280.00 | 432.00 | 1078.00 | 145.80 |
| Co | 4.2 | 2.00 | 2.40 | 1.70 | 1.30 | 2.62 | 4.80 | 1.44 | 2.28 |
| Ni | 9 | 5.86 | 5.52 | 6.42 | 2.90 | 3.58 | 7.30 | 4.88 | 4.38 |
| Cr | 15 | 10.24 | 6.16 | 5.72 | 6.54 | 5.52 | 7.60 | 12.60 | 6.58 |
| Zn | 18 | 15.56 | 9.52 | 60.00 | 10.28 | 10.68 | 12.30 | 13.78 | 9.58 |
| Pb | 6.6 | 3.06 | 4.00 | 2.42 | 2.76 | 3.46 | 7.62 | 3.90 | 4.72 |
| Mg | 200 | 46.40 | 47.00 | 47.20 | 47.20 | 46.80 | 46.60 | 47.20 | 46.80 |
| Cd | 0.24 | 0.52 | 0.64 | 0.64 | 0.74 | 0.62 | 0.50 | 0.92 | 0.74 |
| Na | 200 | 40.00 | 42.80 | 49.20 | 34.40 | 87.60 | 187.60 | 41.60 | 51.60 |
| K | 100 | 7002.00 | 6398.00 | 7812.00 | 6332.00 | 3768.00 | 7172.00 | 6742.00 | 5064.00 |
| Ca | 10000 | 1558.00 | 1264.00 | 492.00 | 364.00 | 2020.00 | 578.00 | 2220.00 | 2300.00 |
| Mn | 100 ± 6 | 8.16 | 13.18 | 12.88 | 12.38 | 14.20 | 37.20 | 17.60 | 49.40 |

*Note* * World average values (in brown coal) of Ca, Fe, Mn, Na, K and Mg are after Valkovic (1983) and Clarke values (in brown coal) of rest of the elements is after Ketris and Yudovich (2009)

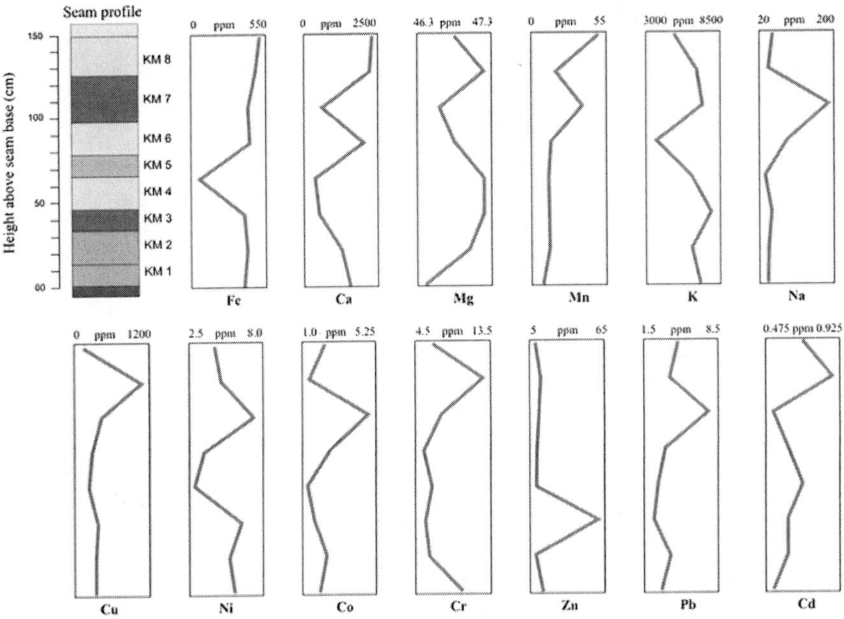


**Figure 10.** Vertical variation of the concentration of major/minor and trace elements along the lignite seam profile

**Table 7.** Concentration of various pyrite forms identified under microscopy in the Kasnau-Matasukh lignites

| No. | Sample number | Disseminated | Discrete | Framboidal | Massive | Fissure |
|---|---|---|---|---|---|---|
| 1 | KM-8 | 25.0 | 37.5 | 6.3 | 18.8 | 12.5 |
| 2 | KM-7 | 11.1 | 22.2 | 44.4 | 11.1 | 11.1 |
| 3 | KM-6 | 46.2 | 30.8 | 23.1 | 0.0 | 0.0 |
| 4 | KM-5 | 35.7 | 21.4 | 21.4 | 14.3 | 7.1 |
| 5 | KM-4 | 28.6 | 28.6 | 14.3 | 0.0 | 28.6 |
| 6 | KM-3 | 50.0 | 16.7 | 16.7 | 16.7 | 0.0 |
| 7 | KM-2 | 33.3 | 33.3 | 0.0 | 33.3 | 0.0 |
| 8 | KM-1 | 100.0 | 0.0 | 0.0 | 0.0 | 0.0 |

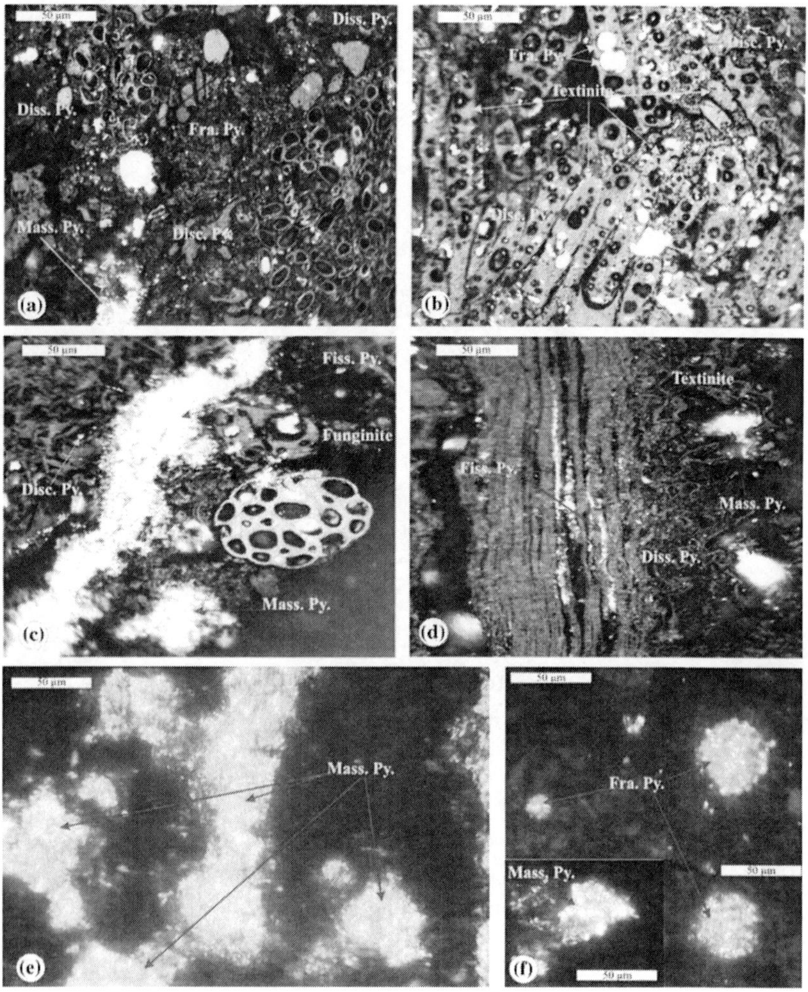


**Figure 11.** Photomicrographs of minerals in the lignite samples of Kasnau-Matasukh; **a** framboidal pyrite, disseminated pyrite, discrete pyrite and massive pyrite; **b** framboidal pyrite and discrete pyrite occuring in textinite maceral; **c** massive pyrite, discrete pyrite, pyrite occurring as fissure and also in funginite maceral; **d** massive pyrite, disseminated pyrite, pyrite occupying the fissures in the lamellar textinite; **e** massive pyrite; **f** massive and framboidal pyrite; abbreviations: *Fra. Py.* framboidal pyrite, *Diss. Py.* disseminated pyrite, *Disc. Py.* discrete pyrite, *Mass. Py.* massive pyrite, *Fiss. Py.* fissure pyrite

As revealed in the correlation matrix among the elements in the Kasnau-Matasukh lignites, Cd shows a strong affinity with Cu while Pb has a strong affinity with Co and Ni. Pb also has an affinity with sulfides especially pyrite. Cr relates strongly with Cu and occurs in sulphides while Co maintains a positive affinity with Na and Mn (Table 8).

**Table 8.** Correlation matrix between minor/trace, group maceral and type of pyrite in Kasnau-Matasukh lignite

| Element | Fe | Ca | Mg | Mn | K | Na | Cu | Co | Ni | Cr | Zn | Pb | Cd | Ash | OM | Humi | Lipti | Inerti |
|---|---|---|---|---|---|---|---|---|---|---|---|---|---|---|---|---|---|---|
| Fe | 1 | | | | | | | | | | | | | | | | | |
| Ca | 0.7 | 1 | | | | | | | | | | | | | | | | |
| Mg | −0.3 | −0.2 | 1 | | | | | | | | | | | | | | | |
| Mn | 0.4 | 0.3 | −0.2 | 1 | | | | | | | | | | | | | | |
| K | −0.2 | −0.6 | 0.14 | −0.2 | 1 | | | | | | | | | | | | | |
| Na | 0.2 | −0.3 | −0.4 | 0.5 | 0.03 | 1 | | | | | | | | | | | | |
| Cu | 0.3 | 0.3 | 0.3 | −0.2 | 0.3 | −0.03 | 1 | | | | | | | | | | | |
| Co | 0.3 | −0.2 | −0.6 | 0.5 | −0.01 | .952** | −0.2 | 1 | | | | | | | | | | |
| Ni | 0.4 | −0.3 | −0.4 | 0.14 | 0.7 | 0.5 | 0.2 | 0.6 | 1 | | | | | | | | | |
| Cr | 0.2 | 0.4 | −0.10 | −0.12 | 0.3 | −0.14 | .805* | −0.2 | 0.2 | 1 | | | | | | | | |
| Zn | 0.04 | −0.4 | 0.3 | −0.3 | 0.5 | −0.13 | 0.02 | −0.2 | 0.4 | −0.2 | 1 | | | | | | | |
| Pb | 0.4 | 0.0 | −0.4 | .708* | 0.03 | .864** | 0.07 | .900** | 0.5 | 0.06 | −0.4 | 1 | | | | | | |
| Cd | −0.02 | 0.4 | .727* | 0.0 | −0.12 | −0.5 | 0.6 | −0.6 | −0.5 | 0.4 | −0.09 | −0.3 | 1 | | | | | |
| Ash | −0.02 | −0.3 | −0.2 | .728* | 0.2 | .744* | −0.03 | 0.6 | 0.3 | 0.001 | −0.11 | .786* | −0.2 | 1 | | | | |
| OM | 0.02 | 0.3 | 0.2 | −.728* | −0.2 | −.744* | 0.03 | −0.6 | −0.3 | 0.001 | 0.11 | −.786* | 0.2 | −1.000** | 1 | | | |
| Humi | −0.15 | 0.12 | −0.15 | 0.6 | −0.3 | −0.09 | −0.6 | −0.09 | −0.4 | −0.3 | −0.09 | −0.01 | 0.06 | 0.4 | −0.4 | 1 | | |
| Lipti | 0.07 | −0.04 | −0.2 | −0.6 | 0.3 | 0.2 | 0.5 | 0.2 | 0.3 | 0.5 | −0.2 | 0.14 | −0.2 | −0.3 | 0.3 | −.904** | 1 | |
| Inerti | 0.2 | −0.2 | .707* | −0.11 | 0.2 | −0.2 | 0.12 | −0.3 | 0.2 | −0.4 | 0.7 | −0.3 | 0.3 | −0.3 | 0.3 | −0.3 | −0.2 | 1 |

*Humi* huminite, *Lipti* liptinite, *Inerti* inertinite, *Py* pyrite, *Diss* disseminated, *Disc* discrete, *Fram* framboidal, *Mass* massive, *Fiss* fissure

* Correlation is significant at the 0.05 level (2-tailed)

** Correlation is significant at the 0.01 level (2-tailed)

# 5 CONCLUSION

1. These lignites are predominantly composed of huminite group of macerals while liptinite and inertinite macerals occur in less concentration. Huminite shows a high concentration at the upper indicating anaerobic degradation during that period. Mineral matter is more towards the upper part of the seam indicating a wet environment.

2. Volatile matter content is high while ash yield is moderate.Sulfur content of these lignites is moderately high. There is increase in volatile matter, carbon and sulphur contents towards the upper part of seam.

3. $S_1$ values are low at the bottom while $S_2$ values are more at the middle part of the seam and decreases towards the top as well as bottom. Total organic carbon content is more in the middle part of the seam and decreases towards the top. Study reveals that these lignites are type-III kerogen and are mainly gas prone.

4. XRD study reveals the presence of mixed clay minerals including kaolinite, illite and chlorite. The peaks in FTIR spectra between 1100 and 400 $cm^{-1}$ further support the presence of these clay minerals.

5. The concentration of Cu is very high in all the samples and over 70 times in KM-7 band. Similarly, Cd is 2–3 times high in almost all the samples while Zn is high in KM-3 band. The concentration of elements like Mn, Na, Cu, Ni, Co, Cr, Pb and Cd is higher towards the upper part of the seam. Preferential enrichment of Ni, Pb, and Co is seen in pyrite.

6. Ash content shows a strong affinity with Mn, among the major elements, and with Co among the trace elements. On the other hand, inertinite maceral has an affinity with Mg and Zn while huminite with Mn, and liptinite with Cu and Cr. Cadmium shows a strong affinity with Mg and Cu while Pb has a strong affinity with Mn, Na, Co and Ni. Chromium relates strongly with Cu,Pb with pyrite, Co with Na and Mn; and Ni with Na, K and Co.

Nevertheless, the results warrant further study for formulating any strategy for proper utilization of the Kasnau-Matasukh lignites of Rajasthan.

## ACKNOWLEDGMENTS

The authors thankfully acknowledge the Department of Geology, Banaras Hindu University for extending the laboratory and other facilities. The help received for Rock–Eval-6 pyrolysis from the R & D department of Oil India Ltd, Duliajan, is thankfully acknowledged. The authors also thank CMPDI, Ranchi for carrying out ultimate analysis. The help rendered by the officials in the Kasnau-Matasukh mine is thankfully acknowledged. We are also thankful to the editor for editorial handling of this manuscript.

## REFERENCES

1. Bertrand PR (1989) Microfacies and petroleum properties of coals as revealed by a study of North Sea Jurassic coals. Int J Coal Geol 13(1–4):575–595

2. BIS (2003) Methods of test for coal and coke (2nd revision of IS: 1350). Part I, Proximate analysis. Bureau of Indian Standard 1–29

3. Blanford WT (1876) On the physical geography of the Great Indian Desert with special reference to the former existence of the sea in the Indus Valley; and on the origin and mode of formation of the sand hills. J Asiat Soc Bengal 45(2):86–103

4. Bouška V, Pešek J, Sýkorová I (2000) Probable modes of occurrence of chemical elements in coal. Acta Mont Ser B 10(117):53–90

5. Cudmore JF (1977) Evaluation of coals for conversion to liquid hydrocarbons. In: International: coal borehole evaluation, Victoria, Australia. The Australasian Institute of Mining and Metallurgy, pp 146–158

6. Dai SF, Ren DY, Tang YG, Yue M, Hao LM (2005) Concentration and distribution of elements in Late Permian coals from western Guizhou Province, China. Int J Coal Geol 61(1–2):119–137

7. Dai SF, Zhang WG, Ward CR, Seredin VV, Hower JC, Li X, Song WJ, Wang XB, Kang H, Zheng LC, Wang PP, Zhou D (2013) Mineralogical and geochemical anomalies of late Permian coals from the Fusui Coalfield, Guangxi Province, southern China: influences of terrigenous materials and hydrothermal fluids. Int J Coal Geol 105:60–84

8. Dai SF, Yang JY, Ward CR, Hower JC, Liu HD, Garrison TM, French D, O'Keefe JMK (2015a) Geochemical and mineralogical evidence for a coal-hosted uranium deposit in the Yili Basin, Xinjiang, northwestern China. Ore Geol Rev 70:1–30

9. Dai SF, Hower JC, Ward CR, Guo WM, Song HJ, O'Keefe JMK, Xie PP, Hood MM, Yan XY (2015b) Elements and phosphorus minerals in the middle Jurassic inertinite-rich coals of the Muli Coalfield on the Tibetan Plateau. Int J Coal Geol 144–145:23–47

10. Dale LS, Chapman JF, Buchanan SJ, Lavrencic SA (1999) Mechanisms for trace element partitioning in Australian coals—Project 4.2 Final report, Co-operative Research Centre for Black Coal Utilisation

11. Davis A, Spackman W, Given PH (1976) The influence of the properties of coals on their conversion to clean fuels. Energy Sources Part A 3:55–81

12. Durand B (1983) Present trends in organic geochemistry in research on migration of hydrocarbons. In: Bjorey M et al (eds) Advances in organic geochemistry 1981. Wiley, Chichester, pp 117–128

13. Eaton AD, Clesceri LS, Greenberg AE (1995) Standard methods for the examination of water and waste water, 19th edn. American Public Health Association, Washington.

14. Eskenzy GM (2009) Trace elements geochemistry of the Dobrudza coal basin, Bulgaria. Int J Coal Geol 78(3):192–200

15. Eskenzy GM, Stefanova YS (2007) Trace elements in the Goze Delchev coal deposit, Bulgaria. Int J Coal Geol 72(3–4):257–267

16. Espitalié J, Laporte JL, Madec M, Marquis F, Leplat P, Poulet J, Boutefeu A (1977) Rapid method of characterizing source rocks and their petroleum potential and degree of maturity. Revue de l'InstitutFrancais du Petrole 32:23–42

17. Espitalié J, Marquis F, Barsony I (1984) Geochemical logging. In: Voorhees KJ (ed) Analytical pyrolysis: techniques and applications. Butterworth, London, pp 276–304

18. Espitalié J, Deroo G, Marquis F (1986) La pyrolyse Rock-Evaletses applications, partie III. Rev Inst Fr Pet 41:73–89

19. Faruqi NH (1978–79) Report on the elucidation of solid geology of sand and alluvium covered areas in part of Bikaner and Ganganagar district, Rajasthan. Geological Survey of India, Government of India. (Unpublished report)

20. Faruqi NH (1982–83) Report on the Geological mapping of pre-quaternaries in parts of Bikaner and Ganganagar district, Rajasthan. Geological Survey of India, Government of India. (Unpublished report)

21. Finkelman RB (1994) Modes of occurrence of potentially hazardous elements in coal: levels of confidence. Fuel Process Technol 39:21–34

22. Finkelman RB (1995) Modes of occurrence of environmentally-sensitive trace elements in coal. In: Swain DJ, Goodarzi F (eds) Environmental aspects of trace elements in coal. Kluwer Academic Publishers, Dordrecht, pp 24–50

23. Georgakopoulos A, Iordanidis A, Karma V (2003) Study of low rank Greek coals using FTIR spectoscopy. Energy Sources A 25:995–1005

24. Ghose KP (1983) Generalised structural and metamorphic map around Batia, Udaipur District, Rajasthan; a study. Indian J Earth Sci 10:228–231

25. Goodarzi F (2002) Mineralogy, elemental composition and modes of occurrence of elements in Candian feed-coals. Fuel 81(9):1191–1213

26. Gowrisankaran S, Sethi PP, Hariharan R, Agrawal, KP (1987) Lignite deposits of India- their occurrences, depositional features and characteristics. In: Singh. R. M. (ed). Proceedings of national seminar coal resources of India. Tara Printing Works, Kamachha, Varanasi, pp 481–553

27. Inan S, Yalcin MN, Mann U (1998) Expulsion of oil from petroleum source rocks: inferences from pyrolysis of samples of unconventional grain size. Org Geochem 29(1–3):45–61

28. International Commitee for Coal and Organic Petrology (2001) The new inertinite classification (I.C.C.P. System 1994). Fuel 80:459–471

29. ISO 11760 (2005) Classification of coals. International standard, pp 1–9

30. ISO 7404-5 (2009) Methods for the petrographic analysis of coals—part 5: method of determining microscopically the reflectance of vitrinite 1–14

31. Jasper K, Hartkopf-Froder C, Flajs G, Littke R (2010) Evolution of Pennsylvanian (Late Carboniferous) peat swamps of the Ruhr Basin, Germany: comparison of palynological, coal petrographical and organic geochemical data. Int J Coal Geol 83(4):346–365

32. Jodha BS (2008) Report on search for lignite by scout drilling in Nagrasar area, Palana basin, Bikaner and Jaisalmer districts, Rajasthan. Geological Survey of India, Government of India

33. Jodha BS (2009) Report on search for lignite by scout drilling in Borana East area, Palana basin, Bikaner and Jaisalmer districts, Rajasthan. Geological Survey of India, Government of India

34. Karr C Jr (1978) Analytical methods for coal and coal products, vol II. Academic Press, New York

35. Ketris MP, Yudovich YE (2009) Estimations of Clarkes for carbonaceous biolithes: world average for trace element contents in black shales and coals. Int J Coal Geol 78(2):135–148

36. Koeverdon JHV, Karlsen DA, Backer-Owe K (2011) Carboniferous non-marine source rocks from Spitsbergen and Bajornoya: coparision with western Arctic. J Pet Geol 34(1):53–66

37. Krevelen V (1961) Coal. Elsevier, Amsterdam, pp 162–210

38. Kulshrestha SK, Singh RY, Atta, Sobeh V (1989) Stratigraphy of lower tertiary sediments in Bikaner, Western Rajasthan. In: Proceedings of XIII Indian Colloq, Micropal. Strat., pp 193–202

39. Lal B, Regar RL (1991) Geophysical investigation for lignite in Sujasar-Kesardesar area, Bikaner district, Rajasthan. Geological Survey of India, Government of India. (Unpublished report)

40. Li Z, Ward CR, Gurba WL (2010) Occurrence of non-mineral inorganic elements in macerals of low-rank coals. Int J Coal Geol 81(4):242–250

41. Lindholm R (1987) A practical approach to sedimentology. Allen & Unwin (New Zealand) Ltd in association with the Port Nicholson Press Ltd 60 Cambridge Terrace, Wellington, pp 124–132

42. Mc Auliffe CD (1979) Oil and gas migration-chemical and physical constraints. Am Assoc Pet Geol Bull 63(5):761–781

43. Mukhopadhyay AK (1974–1975) Report on the elucidation of solid geology of sand and alluvium covered areas in part of Churu and Bikaner district, Rajasthan. Geological Survey of India, Government of India. (Unpublished report)

44. Munshi RL (1975–77) Report on the elucidation of solid geology of sand and alluvium covered areas in part of Bikaner and Ganganagar district, Rajasthan. Geological Survey of India, Government of India. (Unpublished report)

45. Newman J, Price LC, Johnston JH (1997) Hydrocarbon source potential and maturation in Eocene New Zealand vitrinite-rich coals. J Pet Geol 20:137–163

46. Petersen HI (2005) Oil Generation from coal source rocks: the influence of depositional conditions and stratigraphic age. Geol Survey Denmark Greenl Bull 7:9–12

47. Petersen HI, Andsbjerg J, Bojesen-Koefoed JA, Nytoft HP (2000) Coal-generated oil: source rock evaluation and petroleum geochemistry of the Lulita oilfield, Danish North Sea. J Pet Geol 23(1):55–90

48. Pickhardt WV (1989) Trace elements in minerals of German bituminous. Int J Coal Geol 14(1–2):137–153

49. Powell TG (1978) An assessment of the hydrocarbon source rock potential of the Canadian Arctic Islands. Geological Survey of Canada, pp 78–12

50. Powell TG, Boreham CJ (1994) Terrestrial sourced oils: where do they exist and what are our limits of knowledge?—a geochemical perspective. In: Scott AC, Fleet AJ (eds) Coal and coal bearing strata as oil-prone source rocks, vol 77. Geological Society Special Publication, London, pp 11–29

51. Prachiti PK, Manikyamba C, Singh PK, Balram V, Lakshminarayana G, Raju K, Singh MP, Kalpana S, Arora M (2011) Geochemical systematics and precious metal content of the sedimentary horizons of Lower Gondwanas from the Sattupalli coal field, Godavari Valley, India. Int J Coal Geol 88(2–3):83–100

52. Querol X, Alastuey A, Zhuang XG, Hower JC, Lopez-Soler A, Plana F, Zeng RS (2001) Petrology, mineralogy and geochemistry of the Permian and Triassic coals in the Leping area, Jiangxi Province, southeast China. Int J Coal Geol 48(1–2):23–45

53. Ren DY (1996) Mineral matter in coal. In: Han DX (ed) Coal petrology of China. Publishing House of China University of Mining and Technology, Xuzhou, pp 67–77

54. Ren DY, Xu DW, Zhao FH (2004) A preliminary study on the enrichment mechanism and occurrence of hazardous trace elements in the Tertiary lignite from the Shenbei coalfield, China. Int J Coal Geol 57(3–4):187–196

55. Riley KW, French DH, Farrell OP, Wood RA, Huggins FE (2012) Modes of occurrence of trace and minor elements in some Australian coals. Int J Coal Geol 94:214–224

56. Roy AB, Jakhar SR (2002) Geology of Rajasthan (Northwest India) precambrian to recent. Scientific Publishers, Jodhpur, pp 1–421

57. Ryan B, Ledda A (1997) A review of sulphur in coal: with specific reference to the Telkwa deposit, north-western British Columbia. Geological Fieldwork, 22 pp

58. Saikia BK, Boruah RK, Gogoi PK (2007) XRD and FT-IR investigation of sub-bituminous Assam coals. Bull Mater Sci 30(4):421–426

59. Schopf JM (1960) Field description and sampling of coal beds. U.S. Geological Servey Bulletin B 1111, pp 25–69

60. Shah SCD, Kar RK (1971) Palynostratigraphic evolution of the Lowest Eocene sediments of India. In: Proceedings of sem. paleopalynology and Indian strat. Calcutta University Publication Calcutta, pp 255–264

61. Singh PK (2012) Petrological and geochemical considerations to predict oil potential of Rajpardi and Vastan Lignite Deposits of Gujarat, Western India. J Geol Soc India 80:759–770

62. Singh PK, Singh MP (2013) Coal resource of India in context of recent developments in clean coal technologies. In: Sharma PR, Yadav RS, Sharma VN (eds) Interdisciplinary advances in environmental and earth system studies, I edn. RK Books, New Delhi, pp 198–212

63. Singh PK, Singh GP, Naik AS (2010) Petrological considerations for benificiation of Indian coals. J Sci Res 54(1–2):51–60

64. Singh PK, Singh AL, Aniruddha K, Singh MP (2011) A study on removal of selected major elements from Indonesian coal through bacteria: environmental implications. World Acad Sci Eng Technol 75:925–935

65. Singh PK, Singh AL, Aniruddha K, Singh MP (2012) Mixed bacterial consortium as an emerging tool to remove hazardous trace metals from coal. Fuel 102:227–230

66. Singh PK, Singh MP, Singh AK, Arora M, Naik AS (2013) The prediction of the liquefaction behavior of the East Kalimantan Coals of Indonesia: an appraisal through petrography of selected coal samples. Energy Sources Part A 35(18):1728–1740

67. Singh AL, Singh PK, Aniruddha K, Yadav A, Singh MP (2014) Experimental study on demineralization of coal with *Pseudomonas mendocina* strain B6-1 bacteria to obtain clean fuel. Energy Explor Exploit 32(5):831–846

68. Singh PK, Rajak PK, Singh MP, Naik AS, Singh VK, Raju SV (2015a) Environmental geochemistry of selected elements in lignite from Barsingsar and Gurha Mines of Rajasthan, Western India. Geol Soc India 86(1):23–32

69. Singh AL, Singh PK, Singh MP, Kumar A (2015b) Environmentally sensitive major and trace elements in Indonesian coal and their geochemical significance. Energy Sources Part A 37(17):1836–1845

70. Singh PK, Rajak PK, Singh VK, Singh MP, Naik AS, Raju SV (2016) Studies on thermal maturity and hydrocarbon potential of lignites of Bikaner-Nagaur basin, Rajasthan. Energy Explor Exploit 34(1):140–157. doi:10.1177/0144598715623679

71. Swaine DJ (1990) Trace elements in coal. Butterworths & Co, London, p 278

72. Sýkorová I, Pickel W, Christanis M, Wolf K, Taylor GH, Flores D (2005) Classification of huminite. ICCP system 1994. Int J Coal Geol 62:85–106

73. Tang Y, Chang C, Zhang Y, Li W (2009) Migration and distribution of fifteen toxic trace elements during the coal washing of the Kailuan Coalfield, Hebei Province, China. Energy Explor Exploit 27:143–151

74. Taylor GH, Teichmüller M, Davis A, Diessel CFK, Littke R, Robert P (1998) Organic petrology. Gebrüder Borntraeger, Berlin, p 704

75. Tissot BP, Welte DH (1984) Petroleum formation and occurrence, 2nd edn. Springer, Berlin, p 699

76. Turiel JLF, Carvalho WD, Cabanas M, Querol X, Soler AL (1994) Mobility of heavy metals from coal flyash. Environ Geol 23:264–270

77. Valkovic VV (1983) Trace elements in coal, vol 1. CRC Press, Inc, Florida, p 207

78. Wang WF, Qin Y, Sang SX, Zhu YM, Wang CY, Weiss DJ (2008) Geochemistry of rare earth elements in a marine influenced coal and its organic solvent extracts from the Antaibao mining district, Shanxi, China. Int J Coal Geol 76(4):309–317

79. Ward CR (2002) Analysis and significance of mineral matter in coal seams. Int J Coal Geol 50(1–4):135–138

80. Ward CR, Spears DA, Booth CA, Staton I, Gurba LW (1999) Mineral matter and trace elements in coals of the Gunnedah Basin, New South Wales, Australia. Int J Coal Geol 40(4):281–308

81. Ward CR, Matulis CE, Tayler JC, Dale LS (2001) Quantification of mineral matter in the Argonne Premium coals using interactive Rietveld-based X-ray diffraction. Int J Coal Geol 46(2–4):67–82

82. Zodrow EL, Mastalerz M, Werner-Zwanziger U, D'Angelo JA (2010) Medullosalean fusain trunk from the roof rocks of a coal seam: insight from FTIR and NMR (Pennsylvanian Sydney Coalfield, Canada. Int J Coal Geol 82(1–2):116–124

# Chapter 11

# Nanocomposite Sodalite/Ceramic Membrane for Pre-Combustion $CO_2$ Capture: Synthesis and Morphological Characterization

**Michae[l] O. Daramola[1] • Olawale Oloye[1] • Abu Yaya[2]**

[1] *School of Chemical and Metallurgical Engineering, Faculty of Engineering and the Built Environment, University of the Witwatersrand, Wits, Johannesburg 2050, South Africa*

[2] *Department of Materials Science & Engineering, School of Engineering Sciences, University of Ghana, P.O Box LG 77, Legon, Accra, Ghana*

## ABSTRACT

Carbon capture and storage (CCS) is amongst the possible options to reduce $CO_2$ emission. In the application of CCS, $CO_2$ capture techniques such as adsorption and membrane system have been proposed due to less energy requirement and environmental benign than the absorption process. However, membrane system has drawbacks such as poor membrane reproducibility, scale-up difficulty and high cost of the membrane supports. In this study synthesis and characterization of nanocomposite sodalite (HS)/ceramic membrane via "pore-plugging" hydrothermal synthesis (PPH) protocol for pre-combustion $CO_2$ capture is reported. The morphology and crystallinity of the as-prepared membranes were checked with scanning electron microscopy and X-ray diffraction. Surface chemistry of the membrane was examined with Fourier Transform Infrared spectroscopy. In nanocomposite architecture membranes, zeolite crystals are embedded within the pores of the supports instead of forming thin-film layers of the zeolite crystals on the surface of the supports. Compared to the conventional in situ direct hydrothermal synthesis, membranes obtained from PPH possess higher mechanical strength and thermal stability. In addition, defect control with nanocomposite architecture membranes is possible because the zeolite crystals are embedded within the pores of the support, thereby limiting the maximum defect size to the pore size of the support. Furthermore, the nanocomposite architecture nature of the membranes safeguards the membrane from shocks or abrasion that could promote formation of defects. The aforementioned advantages of the nanocomposite architecture membranes could be beneficial in developing high performance and cost-effective membrane materials for pre-combustion $CO_2$ capture.

**Keywords:** Nanocomposite Sodalite Membrane Carbon capture and storage

# 1 INTRODUCTION

Investigations have shown that greenhouse gases (GHG), such as carbon dioxide ($CO_2$), is the cause of global climate change and other related environmental disasters. In sub-Saharan Africa, industrial activities such as coal-fired power plant for the generation of electricity, mining activities, production of cement, coal mining and steel production will generate large quantity of $CO_2$ to the atmosphere. Numerous strategies are being applied today to mitigate continuous emission of GHG, some of which include, energy efficiency in electrical appliances, renewable energy; wind, tidal, solar, water energy, geothermal and biomass, nuclear energy, less carbon intensive sources; natural gas and hydrogen fuels, increase in biological sinks; forestation and carbon sequestration which enables the continuous use of fossil fuels (IPCC 2005). From these available options, carbon sequestration occurs to be a renewable and sustainable form with great potentials. Carbon sequestration mostly known as carbon capture and storage which can best be described as a technology in which $CO_2$ is separated from other gases either from mobile or stationary point source. The captured $CO_2$ is then compressed, transported and stored either in a geological formation, ocean storage or used industrially as a chemical feedstock (Pires et al. 2011).

Amongst the capture approaches, pre-combustion $CO_2$ capture provides opportunity for economic feasibility of the application of membrane system due to the high pressure of $CO_2$ in the gas mixture. In a pre-combustion setup, the fuel (coal) is partially oxidized with oxygen ($O_2$), air or steam to liberate carbon monoxide (CO) and hydrogen ($H_2$) in a chemical process known as reforming. A further shift reaction of carbon monoxide and hydrogen liberates carbon dioxide ($CO_2$) and hydrogen ($H_2$). This occurs in a water gas shift (WGS) reactor where the $CO_2/H_2$ gas mixtures can be separated into the individual components. Hydrogen can serve as fuel in a gas turbine combined cycle plant. This technology is employed in Integrated Gasification Combined Cycle (IGCC) for coal gasification. The $CO_2$ concentration in the IGCC is in the range of 35 %–40 % (vol/vol) with a total pressure of about 20 bar or more in the stream leaving the IGCC, forming a good basis for $CO_2$ pre-combustion capture using a membrane system (Olajire 2010). Furthermore, compared to existing technologies like absorption and adsorption, membranes possess noticeable advantages over absorption and adsorption processes, they have simple modular system, no waste streams and no regeneration energy required. Membrane technology eliminates limitations of conventional gas absorption towers and adsorption processes, thus offering high selectivity and high driving force with a simple modular design (Aoki et al. 2000).

Polymer membranes have been extensively studied for $CO_2$ capture, but polymeric membranes are prone to degradation in harsh conditions, making them unsuitable for $CO_2$ capture from flue gas (with contaminants like sulphur (IV) oxide and water vapour) (Kazemimoghadam and Mohammadi 2005); Chung et al. 2007; Olajire 2010). However, zeolite based membrane technology provides CCS with a great numbers of potentials such as chemical inertness, high thermal stability. For example, sodalite-supported membrane systems possess good chemical resistance and high thermal stability than the polymeric membranes. Zeolite membranes are known to be thermally stable up to 450 °C with reasonable mechanical strength, making them materials of choice in catalytic and separation applications (Khajavi et al. 2007; Daramola et al. 2015; Sunita and Sib 2015).

Out of the proposed technologies for CCS, adsorption and membrane systems have been extensively researched because of less energy consumption and environmental friendliness of these systems. However, reproducibility of highly $CO_2$ selective membranes and high cost of membrane supports (for inorganic membranes) are major setbacks for membrane applications (Daramola et al. 2012). In this study synthesis and characterization of nanocomposite sodalite (HS)/ceramic membrane via "pore-plugging" hydrothermal synthesis (PPH) protocol for pre-combustion $CO_2$ capture is reported for the first time. Compared to the conventional in situ direct hydrothermal synthesis, membranes obtained from PPH possess higher mechanical and thermal stability. In addition, defect control with nanocomposite architecture membrane is possible because the zeolite crystals are embedded within the support pores of the membrane, thereby limiting the maximum defect size to the pore sizes of support. Furthermore, hydroxy sodalite (HS), which is a crystalline zeolite made of cubic array of β-cages as primary building block (Breck 1974), possesses cage aperture in the range of 0.26 and 0.29 nm, making it a good candidate for separation of small gas molecules like water (0.27 nm), helium (0.26 nm) and hydrogen (0.27 nm) under certain process conditions (Khajavi et al. 2007; Daramola et al. 2015). In addition, the nanocomposite architecture nature of the membrane, which could protect the separative layer of the membrane from abrasion during membrane assembly, could be instrumental to developing high performance $H_2$-selective membrane for IGCC system.

# 2 EXPERIMENTAL PROCEDURE

## 2.1 Material and method

The chemicals used for the synthesis of the membranes, anhydrous sodium metasilicate ($Na_2SiO_3$), sodium hydroxide (NaOH) and anhydrous sodium aluminate ($NaAlO_2$) were purchased from Sigma Aldrich (Pty) SA. The de-ionized water was prepared in-house and the chemicals were used as supplied without any further purification. The substrates for the membranes were asymmetrical α-alumina tubes (o.d. 10 mm, i.d. 7 mm, length 5 cm) purchased from Pall-Exekia (Germany). The cross-sectional layers of the support were of dimensions: outer layer, 1200 nm; intermediate layer, 800 nm; inner layer, 200 nm.

## 2.2 Membrane synthesis and characterization

Prior to membrane synthesis, the supports were cleaned in boiling water for 1–2 h to remove dirt and loose particles from their pores. The synthesis of the membrane was carried out using pore-plugging technique as described in Daramola et al. (2015) and illustrated schematically in Fig. 1. Certain quantities of anhydrous sodium aluminate ($NaAlO_2$), anhydrous sodium metasilicate ($Na_2SiO_3$), sodium hydroxide and de-ionised water were measured and vigorously mixed in a polytetrafluoroethylene bottle on a magnetic stirrer. About 45 mL of the homogeneous mixture was then poured into a Teflon-lined stainless steel autoclave and the cleaned support was then inserted into the solution in the autoclave and sealed. In order to obtain nanocomposite architecture membrane via pore plugging hydrothermal synthesis, the sealed autoclave was subjected to hydrothermal synthesis with interruption at 413 K using the temperature programme described in

Daramola et al. (2015). The morphology of the cleaned support, the as-prepared membranes and the sodalite particles was checked with scanning electron microscopy (SEM) (Phillips XL 20). Furthermore, the crystal phases in the as-prepared samples were identified with X-ray diffraction (XRD) (model: Bruker D8 advance X-ray diffractometer) using CoKα radiation (λ = 0.179 nm) at a scan rate of 0.25 s/step and a step size of 0.02°. Fourier transform infrared (FTIR) analysis was conducted with Bruker IFS spectrometer using KBr pellets as background to obtain infrared spectrum that confirmed the purity of the sodalite crystals embedded within the pores.

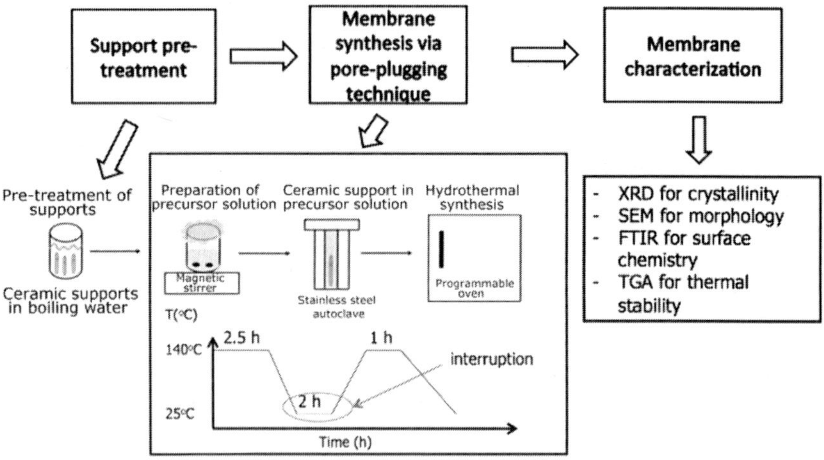


**Figure 1.** Experimental procedure for the synthesis of the sodalite/ceramic membranes

# 3 RESULTS AND DISCUSSION

## 3.1 Characterization of support and HS crystals

The support was characterized before the synthesis. In addition, the HS crytsals ontained from the bottom of the autoclave after the PPH were characetrized as well. The SEM micrographs showing the cross-sectional view of the support is depicted in Fig. 2. The SEM micrographs, the FTIR spectrum and the XRD pattern of the HS crystals obtained from the bottom of the Teflon-lined autoclave are presented in Fig. 3a–c. The SEM micrograph of the HS crystals depicted in Fig. 3a shows cubic and nanorod-like shape of the crystals. Literature has confirmed other shapes of hydroxy sodalite such as flower-like, urchin, thread-ball and cubic/octahedral shapes (Bayati et al. 2008; Kundu et al. 2010; Naskar et al. 2011), making the results from this study a re-affirmation of these findings. The formation of other morphological shapes, as observed in this study, could be attributed to the interplay between the growth process and the nucleation process that seems to proceed in parallel once high temperature is applied during the synthesis (Chen et al. 2007).

The FTIR spectrum in Fig. 3b shows the chemical nature of the HS crystals. The various functional groups obtained from the FTIR are presented and explianed in Table 1. This information from the infrared spectrum of the HS crystals is consistent with literature (Naskar et al. 2011; Buhl et al. 2011). The XRD pattern of the HS crystals (see Fig. 3c) confirm that pure sodalite crystals were obtained during the synthesis.

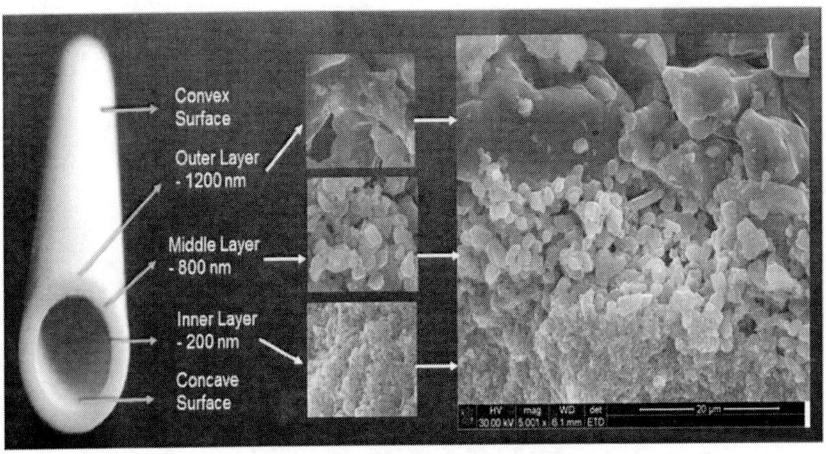


**Figure 2.** SEM micrograph of the cross-sectional view of the membrane support

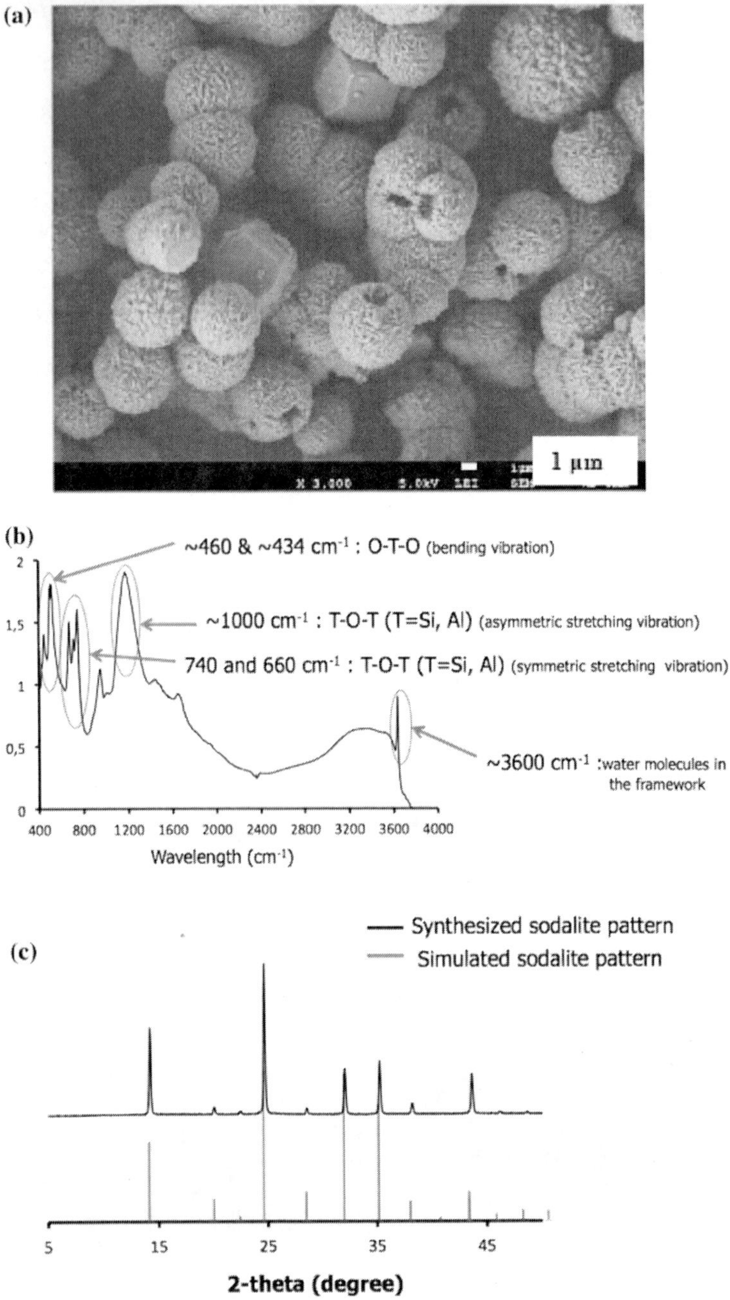


**Figure 3.** Morphology, crystallinity and chemical nature of the sodalite particles. SEM micrograph (**a**), FTIR spectrum (**b**), and XRD pattern (**c**)

**Table 1.** Functional groups from FTIR and interpretation

| Functional group | Wavelength position | Attribute |
|---|---|---|
| T–O–T (T=Si, Al) | 1000 $cm^{-1}$ | Asymmetric stretching vibration of T–O–T |
| T–O–T (T=Si, Al) | Around 740 and 660 $cm^{-1}$ | Symmetric stretching vibration of T–O–T |
| Hydroxyl (–OH) | 3600 $cm^{-1}$ | Presence of water molecule forming part of the case structure |

## 3.2 Characterization of the as-prepared nanocomposite HS/ceramic membrane

The morphology of the cross-sectional view and the surface view of the synthesized membrane are depicted in Fig. 4(top, middle, and bottom). Sodalite crystals could be seen vividly in all the SEM micrographs depicted in Fig. 4(top, middle, and bottom). It is clearly shown that the 200 nm layer of the support seems totally pore-plugged with sodalite crystals (see Fig. 4(middle)). Furthermore, the 800 nm layer of the support contains a lot of the sodalite crystals but not totally plugged when compared to the 200 nm layer. Also the 1200 nm layer of the support has a few crystals embedded within it. The total pore-plugging of the 200 nm layer indicates that the membrane might be defect free, but evaluation of te quality of the membrane using methods such as quick permeation test and/or gas mixture permeation will be required to confirm this speculation. The formation of the nanocomposite architecture could be attributed to the insitu heterogeneous nucleation during the synthesis. As described elsewhere (Li et al. 2008; Daramola et al. 2009, 2015), the HS nanocrystals (could be called seeds as well) formed in situ during the first stage of the PPH (before interruption) of the synthesis promote heterogenous nucleation and growth during the second stage (after interruption) of the synthesis. Eventually, the heterogenous nucleation during the post-interruption stage enhanced the growth into bigger crystals. Change in autogenous pressure in the vessel during the interruption stage, as a result of the change in temperature, might have enhanced the flow of more precursor solution into the porous structure of the support, thereby resulting in the growth of HS crystals within the pores that pore-plugged the pores of the support. The difference between this method and the secondary seeded growth (SSG) technique is in the in situ formation of the seeds insteading of ex situ formation of seeds as the case is in SSG. The ex situ formation of seeds in SSG makes the technique time demanding than the PPH used in this study and SSG technique always produces “thin-film” membranes.

**Figure 4.** SEM micrographs of the as-prepared nanocomposite membrane. Showing the crosssectional view with the presence of sodalite crystals (*top*), 1200 nm layer of the support with sodalite crystals within it (*middle*), and the surface view of the membrane (*bottom*)

### 3.3 Thermal stability of the as-prepared nanocomposite HS/ceramic membranes

A "thin-film" HS membrane was prepared as described elswehere (Khajavi et al. 2007) using similar membrane support for comparison. The thermal stability of the as-prepared thin-film membranes was compared to that of the nanocomposite membranes. The thermal stability was investigated by observing the thermal decomposition of a small portion cut-off from the membrane in a TGA. The thermal stability of the nanocomposite sodalite membrane compared to the "thin-film" sodalite membrane and the HS crystals is depicted in Fig. 5. It is vivid from Fig. 5 that the two membranes are more thermal stable when compared to the thermal stability of the HS crsystals. However, the nanocomposite membrane displayed a higher thermal stability than that of the "thin-film" counterpart. The enhanced thermal stability of the nanocomposite sodalite membrane can be attributed to the nanocomposite architecture nature of the membrane (the HS crystals are embedded within the pores of the supports instead of having them on the surface of the support).

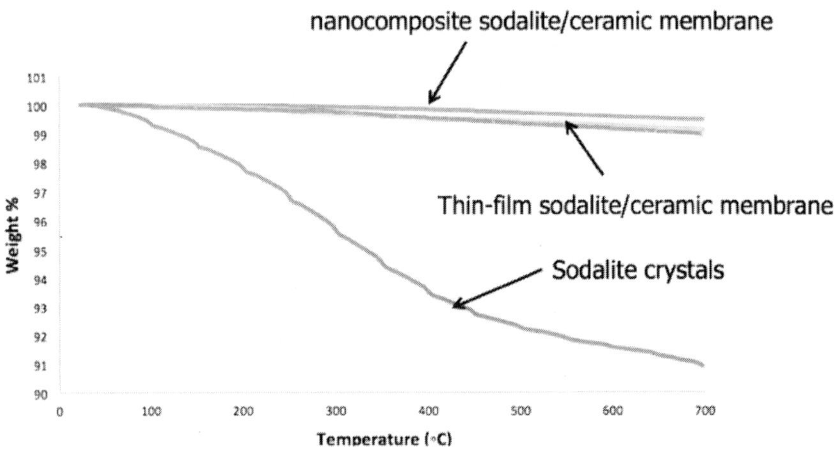


**Figure 5.** Thermal stability of the nanocomposite membrane compared to the thin-film HS membrane and HS crystals

# 4 CONCLUSIONS AND RECOMMENDATIONS

In this article, preparation and morphological characterization of nanocomposite sodalite/ceramic membranes obtained via the pore-plugging hydrothermal synthesis technique have been documented. The SEM micrographs of the as-prepared membranes reveal that nanocomposite sodalite/α-alumina membranes contain sodalite crystals embedded within the pore structure of the support. In addition, the nanocomposite membrane displayed higher thermal stability than that of the "thin-film" membrane synthesized via the conventional direct hydrothermal method. The membrane could be employed to separate $CO_2$ from $H_2/CO_2$ mixture in order to capture the $CO_2$ gas from IGCC. As far as could be ascertained, this article presents the first open report on the preparation and morphological characterization of nanocomposite sodalite/ceramic membranes for pre-combustion $CO_2$ capture. However dynamic tests, such as single pas permeation test and $CO_2/H_2$ separation test should be carried out on the membranes to evaluate the separation performance of the membranes during pre-combustion $CO_2$ capture and at pre-combustion $CO_2$ capture conditions. Research effort is ongoing in our laboratory on the aforementioned tests and the results obtained thereof will be communicated in the near future.

# ACKNOWLEDGMENTS

MOD acknowledges the assistance of Ameera and Sameera on the synthesis of the membrane samples.

# REFERENCES

1. Aoki K, Kusakabe K, Morooka S (2000) Separation of gases with an A-type zeolite membrane. Ind Eng Chem Res 39(7):2245–2251

2. Bayati B, Babaluo AA, Karimi R (2008) Hydrothermal synthesis of nanostructure NaA zeolite: the effect of synthesis parameters on zeolite seed size and crystallinity. J Eur Ceram Soc 28(14):2653–2657

3. Breck DW (1974) Zeolite molecular sieves: structure, chemistry and use. Wiley, New York

4. Buhl JC, Schuster K, Robben L (2011) Nanocrystalline sodalite grown from superalkaline NaCl bearing gels at low temperature (333 K) and the influence of TEA on crystallization process. Microporous Mesoporous Mater 142:666–671

5. Chen X, Qiao M, Xie S, Fan K, Zhou W, He H (2007) Self-construction of core-shell and hollow zeolite analcime icositetrahedra: a reversed crystal growth process via oriented aggregation of nanocrystallites and recrystallization from surface to core. J Am Chem Soc 129(43):13305–13312

6. Chung TS, Jiang YL, Yi L, Kulprathipanja S (2007) Mixed matrix membranes (MMMs) comprising organic polymers with dispersed inorganic fillers for gas separation. Prog Polym Sci 32:483–507

7. Daramola MO, Burger AJ, Pera-Titus M, Giroir-Fendler A, Miachon S, Lorenzen L, Dalmon JA (2009) Nanocomposite MFI–ceramic hollow fibre membranes via pore-plugging synthesis: prospects for xylene isomer separation. J Membr Sci 337(1–2):106–112

8. Daramola MO, Aransiola EF, Ojumu TV (2012) Potential applications of zeolite membranes in reaction coupling separation processes. Materials 5(11):2101–2136

9. Daramola MO, Dinat A, Hasrod S (2015) Synthesis and characterization of nanocomposite hydroxy-sodalite/ceramic

membrane via pore-plugging hydrothermal synthesis technique. J Membr Sep Technol 4(1):1–7

10. Intergovernmental Panel on Climate Change (2005) Special report on carbon dioxide capture and storage. Cambridge University Press, Cambridge

11. Kazemimoghadam M, Mohammadi T (2005) Separation of water/UDMH mixtures using hydroxysodalite zeolite membranes. Desalination 181(1–3):1–7

12. Khajavi S, Kapteijn F, Jansen CJ (2007) Synthesis of thin defect-free hydroxy sodalite membranes: new candidate for activated water permeation. J Membr Sci 299(1–2):63–72

13. Kundu D, Dey B, Naskar MK, Chatterjee M (2010) Emulsion-derived urchin-shaped hydroxy sodalite particles. Mater Lett 64:1630–1633

14. Li Y, Pera-Titus M, Xiong G, Yang W, Landrivon E, Miachon S, Dalmon JA (2008) Nanocomposite MFI-alumina membranes via pore-plugging synthesis: genesis of the zeolite material. J Membr Sci 325:973–981

15. Naskar MK, Kundu D, Chatterjee M (2011) Effect of process parameters on surfactant-based synthesis of hydroxy sodalite particles. Mater Lett 65(3):436–438

16. Olajire AA (2010) $CO_2$ capture and separation technologies for end-of-pipe applications—a review. Energy 35(6):2610–2628

17. Pires JCM, Martins FG, Alvim-Ferraz MCM, Simoes M (2011) Recent developments on carbon capture and storage: an overview. Chem Eng Res Des 89(9):1446–1460

18. Sunita S, Sib KG (2015) Hydrogen the future transportationfuel: from production to applications. Renew Sustain Energy 43:1151–1158

# Chapter 12

# The Properties and Suitability of Various Biomass/Coal Blends for Co-Gasification Purposes

**Sibongiseni Gaqa[1,2], Sampson Mamphweli[2], David Katwire[1], Edson Meyer[2]**

[1]*Chemistry Department, University of Fort Hare, Alice, South Africa*

[2]*Institute of Technology, University of Fort Hare, Alice, South Africa*

## ABSTRACT

Gasification is a promising technology for the production of gaseous fuels, mainly syngas, which is produced from the hydrocarbon-based materials such as coal and biomass. Currently, coal is the main feedstock that is used for the gasification process due to its large reserves and higher energy per volume. However, the use of coal has been a more concern because of the environmental impacts caused by the emission of toxic gases such as the sulphides, sulphates and nitrates as well as the ash slagging problems forming inside the gasifier. On the other hand, biomass is a renewable energy resource of interest as a replacement for coal to reduce the environmental impacts associated with fossil fuel usage. Much consumption of fossil fuels has caused serious energy crisis and environmental impacts, globally. Co-gasification of coal and biomass is considered as a connection between energy production based on fossil fuels and energy production based on renewable fuels. The utilization of biomass by co-gasification with coal causes reductions of carbon dioxide, nitrogen and sulfur emissions due to the renewable character of biomass and low contamination content in biomass. This study determined the properties of various biomass/coal blends and their suitability for co-gasification in a downdraft biomass gasifier. A bomb calorimeter was used to determine the calorific values of the material. CHNS and XRF analysis were carried out to determine the elemental analysis of the material. Thermogravimetric analysis (TGA) was conducted to investigate the thermal degradation of the material. The kinetic analysis of the various feedstocks allows the prediction of the rate at which co-gasification takes place. The results suggested that blending coal with biomass result in a faster reaction rate at lower temperatures than that of coal alone and lower activation energy due to the high quantity of volatile matter in biomass.

**Keywords:** Gasification, Functional Groups, Thermochemical Process, Coal, Biomass

## 1. INTRODUCTION

Energy plays an important role in the world's present and future. Global annual energy demand is constantly and rapidly growing. The best way of providing for this enormous increase is energy diversification, such as declining reserves and fluctuating prices of fossil fuels, national energy security, minimization of environmental impact given the increase in atmospheric carbon dioxide ($CO_2$), and lead to sustainable fuels. In this perspective biomass appears as an attractive resource, potentially to progressively replace fossil fuels [1] . The use of biomass as an alternative energy source can be traced back to the first wood-fuelled fire. However, biomass produces cleaner syngas than coal because of its lower sulphur and ash contents and neutral traces of carbon. There is also a greater degree of complexity involved in processing biomass before gasification starts [2] . Biomass is a world-wide available renewable energy resource. Biomass are a world-wide available renewable energy resource, when produced by sustainable means it removes just about the same amount of carbon dioxide during its energy conversion as that adsorbed by photosynthesis during plant life, as a result, the use of biomass does not contribute to the increase of $CO_2$ in the atmosphere, and this makes it a remarkable energy source in the present context of environmental attention towards the emission levels of greenhouse gases [3] . During gasification, slight oxidation and major pyrolysis take place. During pyrolysis biomass/coal produces char condensable volatiles and tar, and gaseous products. Char is mostly carbon and it follows reactions 1 to 6 (below) during gasification:

Oxidation reaction

$$C + O_2 \leftrightarrow CO_2 \left(+393\,\mathrm{MJ/kg/mole}\right) \tag{1}$$

Methanation reaction

$$C + 2H_2 \leftrightarrow CH_4 \left(+75\ \mathrm{MJ/kg\ mole}\right) \tag{2}$$

Water-gas shift reaction

$$CO + H_2O \leftrightarrow CO_2 + H_2 \left(+122.6\ \mathrm{MJ/kg\ mole}\right) \tag{3}$$

Steam reforming reaction

$$H_2 + 1/2\,O_2 \leftrightarrow H_2O\ \left(+241.1\ \mathrm{kJ/mol}\right) \tag{4}$$

Water gas reaction

$$C + H_2O \leftrightarrow CO + H_2 \left(+42.3\ \mathrm{MJ/kg/mole}\right) \tag{5}$$

Boudouard reaction

$$C + CO_2 \leftrightarrow 2CO(-164.9\ \text{MJ/kg/mole}) \tag{6}$$

All of the volatiles and some tars are thermally broken down into simple gaseous products during gasification. The remaining tar and some other alkali minerals leave with the product gases [4] . A number of studies on the thermo chemical conversion of biomass or residues from wood products have been carried out. These studies revealed in principle a similar behaviour to that of coal. In general, a total gasification process can be generally separated into two main reaction stages called pyrolysis and gasification. The pyrolysis stage corresponds to actions related to devolatilization or thermal decomposition that occur during the initial heating-up of a solid fuel, and is completed only in seconds, or less at high temperatures [5] . Gasification begins with an endothermic process known as pyrolysis, where the small amount of carbon-based feedstock is burned; giving heat that is essential to drive out moisture and volatiles, and all this takes place in the absence or poor presence of oxygen. After pyrolysis, more heat is then needed to thermally break the volatiles and the long hydrocarbon chains into lighter gases. The remaining carbon left in the feedstock is then gasified giving syngas as the product. The chemical form of syngas usually consists mostly of CO, $H_2$ and $CH_4$ as fuel and also $CO_2$, $N_2$, and water vapour as non-combustible gases [2] . The syngas also consists of $H_2S$, COS, HCN, HCl, Hg, and other contaminants that will have to be removed before utilizing the syngas for power generation. Unlike combustion, the aim of gasification is to produce syngas that can then be separated into components and used for different applications in a more efficient way. All of this occurs inside the gasifier. After the volatiles leave the feedstock, what is left in the gasifier is char, which is basically pure carbon. Once this occurs, then the char is reacted using air/oxygen and steam to produce carbon monoxide (CO), carbon dioxide ($CO_2$), and hydrogen ($H_2$) [Long III and Wang, 2011]. Figure 1 shows the basic step wise processing for the co-gasification of coal and biomass:

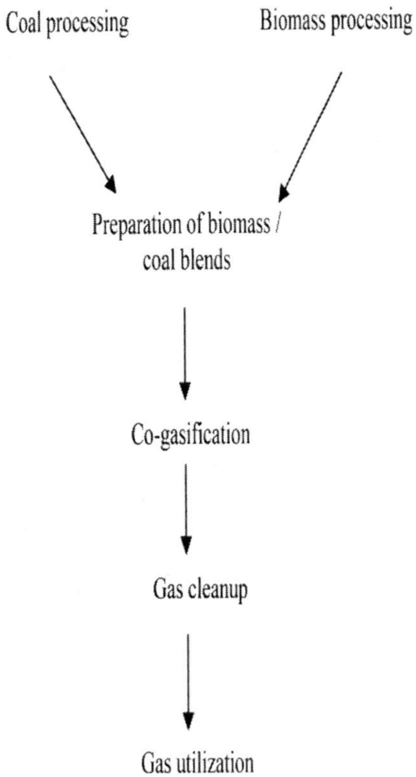


**Figure 1.** Basic step wise processing for the co-gasification of coal and biomass.

## 2. MATERIALS AND METHODS

Samples of eucalyptus and coal chips were used for this study. The cone-and-quaternary method was applied subsequent to size reduction to obtain more representative samples. Chips were milled using a Condux-Werk Wolfgang bei Hanau mill equipped with a 1 mm circular blade to reduce the size of the chips to approximately ±425 μm, the chips were further milled to approximately ±250 μm. To obtain the fractions between 250 and 425 μm the samples were screened in a Retsch AS200 shaker, where the 250 μm was used for these analyses. For quantitative elemental analysis of the materials a CHNS and XRF analysers were used. Thermogravimetric Analyser (TGA) was used to observe the thermal degradation of the material. The methodology employed is outlined in Figure 2.

# 3. RESULTS AND DISCUSSION

## 3.1. Calorific Values

Table 1 shows the energy values of coal, eucalyptus and the coal/eucalyptus blends. This data was obtained using ECO CAL-2K bomb calorimeter, to establish the relation when coal is blended with biomass, whether the energy values of the blends would decrease or increase since the good quality gas also depends on the energy values of the material used for the gas production From Table 1 it was observed that the coal has a lower energy value of about 17.78 MJ/kg which is far less than the amount found in high grade coals of more than 24 MJ/kg [Ganesh, 2006], however this value is acceptable since it is more than 15 MJ/kg reported to be the lowest calorific value of low grade coal. The eucalyptus was observed to have the energy value of about 18 MJ/kg, which is for a typical biomass Lignocellulosic material. The blends showed no significant difference between from the pure coal and pure eucalyptus wood. Although the calorific values of the materials did not differ significantly, the rate of thermal degradation and energy released was different due to the high carbon content in coal and the blends with higher carbon content. This is further discussed during the analysis of thermal degradation (TG).

## 3.2. Proximate Analysis

Table 2 presents the proximate analysis of coal, eucalyptus and the eucalyptus/coal blends.

According to Ganesh, 2006 a typical biomass material has volatile matter content in the range from 60% to 90% and a typical with coal

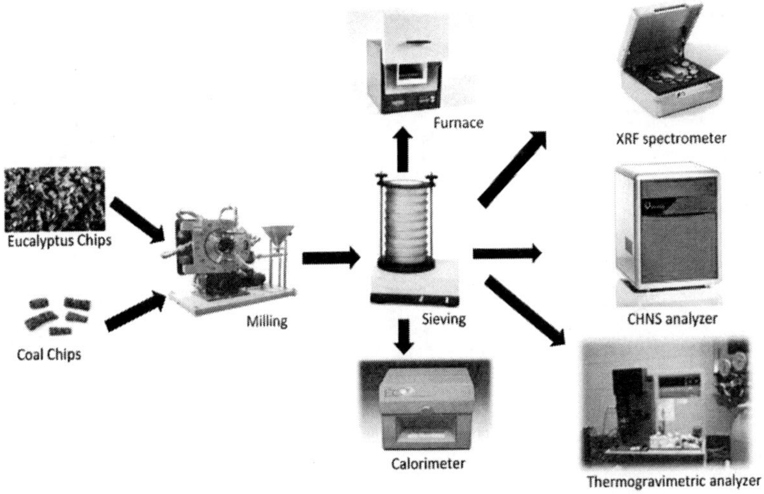


**Figure 2.** Methods employed for sample analysis.

**Table 1.** Calorific values of coal eucalyptus wood and the coal/eucalyptus blends.

| Sample ID | Calorific value (MJ/Kg) |
|---|---|
| Coal | 17.78 |
| Eucalyptus | 18.72 |
| $Euc_{10\%}$-$Coal_{90\%}$ | 17.74 |
| $Euc_{20\%}$-$Coal_{80\%}$ | 17.21 |
| $Euc_{30\%}$-$Coal_{70\%}$ | 17.91 |
| $Euc_{40\%}$-$Coal_{60\%}$ | 18.03 |
| $Euc_{50\%}$-$Coal_{50\%}$ | 17.85 |
| $Euc_{60\%}$-$Coal_{40\%}$ | 17.86 |
| $Euc_{70\%}$-$Coal_{30\%}$ | 17.98 |
| $Euc_{80\%}$-$Coal_{20\%}$ | 17.86 |
| $Euc_{90\%}$-$Coal_{10\%}$ | 17.64 |

Euc: eucalyptus.

**Table 2.** Proximate analysis of coal, eucalyptus and the eucalyptus/coal blends.

| Sample ID | % VM | % FC | % Ash |
|---|---|---|---|
| Coal | 24.27 | 25.94 | 47.48 |
| Eucalyptus | 78.65 | 18.51 | 0.775 |
| $Euc_{10\%}$-$Coal_{90\%}$ | 23.31 | 23.96 | 42.50 |
| $Euc_{20\%}$-$Coal_{80\%}$ | 29.54 | 23.56 | 34.37 |
| $Euc_{30\%}$-$Coal_{70\%}$ | 37.54 | 22.00 | 35.28 |
| $Euc_{40\%}$-$Coal_{60\%}$ | 43.59 | 22.76 | 27.42 |
| $Euc_{50\%}$-$Coal_{50\%}$ | 50.48 | 20.52 | 20.78 |
| $Euc_{60\%}$-$Coal_{40\%}$ | 55.22 | 20.33 | 16.81 |
| $Euc_{70\%}$-$Coal_{30\%}$ | 61.63 | 19.83 | 13.59 |
| $Euc_{80\%}$-$Coal_{20\%}$ | 69.29 | 19.32 | 8.535 |
| $Euc_{90\%}$-$Coal_{10\%}$ | 74.32 | 18.53 | 6.599 |

Euc: eucalyptus.

having volatile matter ranging from 20% to 50%. From Table 2 the volatile matter of the coal used for this study is 24.27% and the volatile matter of the eucalyptus is 78.75%, both these values are within the acceptable ranges. Generally the volatile matter is a mixture of short and long chain hydrocarbons, aromatic hydrocarbons and some sulfur. The bonds between these hydrocarbons are strong in coal and medium to light in the biomass materials, which explains the high volatile matter content in eucalyptus and low content in coal. It is also observed that the blends with the high content of coal have low volatile matter content and the blends with high eucalyptus content have high volatile matter content, which then implies that there is a synergy that exists between eucalyptus and coal during thermochemical conversion, and this synergy is further explained in the TGA section.

The fixed carbon content of a certain material is the carbon that is left after the volatiles have been driven off. It differs from the ultimate carbon content of the coal because some carbon is lost in hydrocarbons with the volatiles. The fixed-carbon value is one of the parameters used in conducting computer simulations. The percentage amount of ash also showed that the coal used for this study is a low grade coal with very high ash content of 47.73%. The eucalyptus on the other hand has low ash content of 0.775%. This implies that this type of coal is not good for gasification because it can cause blockages, slagging in the rector and excessive tar formation depending on its chemical composition. This high ash content in coal is also the reason for the low calorific value of the coal [6] . The blending of this type of coal with eucalyptus could thus lead to the low ash content in the blend and reduce some of the latter risks, blending this type of coal with eucalyptus reduces the ash content. The more eucalyptus added the lower the ash content. Adding 50% eucalyptus to the coal reduces the ash from 42% to 20% implying a 50% decrease in ash content. It is well known that there is a correlation between the ash content and the calorific value of the fuel. Lower ash content should imply higher calorific value; however the results recorded in this study do not suggest a change in the calorific value after blending.

### 3.3. Ultimate Analysis

Table 3 presents the elemental analysis of coal, eucalyptus and the eucalyptus/coal blends. This data was obtained using a Vario EL CUBE CHNS analyzer.

Generally, compared with solid fossil fuels, biomass materials contain much less carbon content and more oxygen [7] . From Table 3 it can observed that the quantity of carbon in coal was found to be higher than that of eucalyptus as expected (about 55.90%). The carbon content of all the eucalyptus/coal blends decreases with an increase in the quantity of eucalyptus in the blend. This is because of some dilution effect caused by the low carbon content in eucalyptus. This may have an impact on the energy value of the feedstock; however, it might not be a challenge as it may be complemented by the reactivity of eucalyptus that would act as a catalyst for full conversion of coal, thereby resulting in more energy released, this can be clearly observed when a high grade coal is used. During combustion sulphur quantitavely oxides and some of it reacts with alkali materials to form sulphates. Alkali sulphates are unstable at typical combustion temperatures of about 900˚C. These sulphates can stimulate agglomeration [6] . As can be observed from Table 3, both coal and eucalyptus result in blends with minimal amounts of sulphur.

**Table 3.** CHNS analysis of coal, eucalyptus and eucalyptus/coal blends.

| Sample ID | %C | %H | %N | %S | %O |
|---|---|---|---|---|---|
| Coal | 55.90 | 3.900 | 1.250 | 0.400 | 38.55 |
| Eucalyptus | 45.30 | 5.500 | 0.070 | 1.000 | 48.13 |
| $Euc_{10\%}$-$Coal_{90\%}$ | 53.39 | 3.980 | 1.150 | 0.340 | 41.14 |
| $Euc_{20\%}$-$Coal_{80\%}$ | 53.01 | 4.100 | 1.120 | 0.398 | 41.37 |
| $Euc_{30\%}$-$Coal_{70\%}$ | 51.49 | 4.000 | 1.100 | 0.465 | 42.94 |
| $Euc_{40\%}$-$Coal_{60\%}$ | 52.26 | 4.250 | 0.980 | 0.473 | 42.04 |
| $Euc_{50\%}$-$Coal_{50\%}$ | 51.03 | 4.429 | 0.850 | 0.500 | 43.19 |
| $Euc_{60\%}$-$Coal_{40\%}$ | 49.69 | 4.320 | 0.543 | 0.702 | 44.74 |
| $Euc_{70\%}$-$Coal_{30\%}$ | 48.82 | 5.201 | 0.220 | 0.798 | 44.96 |
| $Euc_{80\%}$-$Coal_{20\%}$ | 47.55 | 4.730 | 0.099 | 0.990 | 46.63 |
| $Euc_{90\%}$-$Coal_{10\%}$ | 46.71 | 4.900 | 0.089 | 0.975 | 47.33 |

Euc: eucalyptus.

**Table 4** . ANOVA one way test.

| *Groups* | *Count* | *Sum* | *Average* | *Variance* |
|---|---|---|---|---|
| 100%coal | 5291 | 324.928.8 | 61.4116 | 1450.958 |
| 100%euc | 2655 | 202.252.9 | 76.17812 | 559.9832 |
| $Euc_{10\%}$-$coal_{90\%}$ | 2655 | 211.696.3 | 79.73494 | 309.1994 |
| $Euc_{20\%}$-$coal_{80\%}$ | 2655 | 191.909.1 | 72.28214 | 448.6637 |
| $Euc_{40\%}$-$coal_{60\%}$ | 2655 | 170.741.2 | 64.30929 | 676.4141 |
| $Euc_{60\%}$-$coal_{40\%}$ | 1640 | 125.490.8 | 76.5188 | 584.7107 |
| $Euc_{70\%}$-$coal_{30\%}$ | 1640 | 122.984.1 | 74.99028 | 606.9891 |
| $Euc_{80\%}$-$coal_{20\%}$ | 1640 | 118.116.3 | 72.02214 | 728.5802 |
| $Euc_{90\%}$-$coal_{10\%}$ | 1640 | 116.588.7 | 71.09069 | 828.5033 |

ANOVA

| *Source of Variation* | *SS* | *df* | *MS* | *F* | *P-value* | *F crit* |
|---|---|---|---|---|---|---|
| Between Groups | 956.083.7 | 8 | 119.510.5 | 153.6287 | $4.8477 \times 10^{-253}$ | 1.938824 |
| Within Groups | 17.473.589 | 22.462 | 777.9178 | | | |
| Total | 18.429.673 | 22.470 | | | | |

### 3.4. TG and DTG Analysis

Figure 3 and Figure 4 presents the mass loss and derivative mass loss of coal and eucalyptus with respect to temperature and time, respectively. The data was obtained by analysing the samples using Perkin Elmer TGA 4000. All samples were analyzed by purging with Nitrogen at a flow rate of 20 mL/min. All the samples were heated from 35°C to 900°C at 20°C/min. This data was necessary to establish the rate of thermal degradation of coal, eucalyptus and the eucalyptus/coal blends to obtain the basis for comparison of the data on the blends. Lignocellulosic biomass consists of cellulose, hemicellulose and lignin.

Cellulose is the major structural component of plants, especially of wood and plant fibres. It is a linear homogeneous polymer and degrades at higher temperatures (around 300°C - 900°C). Hemicellulose is a polysaccharide found in plant cell walls and it normally decomposes at temperatures above 160°C up to around 300°C. Lignin is found in all vascular plants, and represents about 25% of the terrestrial plant biomass. It is a complex phenolic heteropolymer that results from oxidative polymerization [8] . It is normally lost at a slow rate over a wide temperature rate of 200°C - 500°C.

During thermal degradation (Figure 3) there were three most distinct stages that were observed. In the first stage (A) moisture is released followed by the release of the volatiles in the second stage of degradation (B) and then fixed carbon in the third stage of degradation (C). The coal used had very low moisture content as compared to the biomass materials, but it was found to degrade very slowly due to the presence of long strong carbon - carbon chains as well as the aromatic rings as observed with FT-IR analysis. It was observed that eucalyptus degraded at lower temperatures due to the high quantity of volatile matter content in biomass materials as compared to coal. From the derivative weight loss graph which is Figure 4 it was observed that the rapiddevolatization of coal took place at higher temperature than that of eucalyptus, which was 530°C and 322°C respectively. ThIs implies that coal requires higher gasification temperature than biomass for complete gasification to take place. Figure 5 and Figure 6 presents the mass loss and derivative mass loss of the eucalyptus/coal blends, respectively in relation to temperature and time.

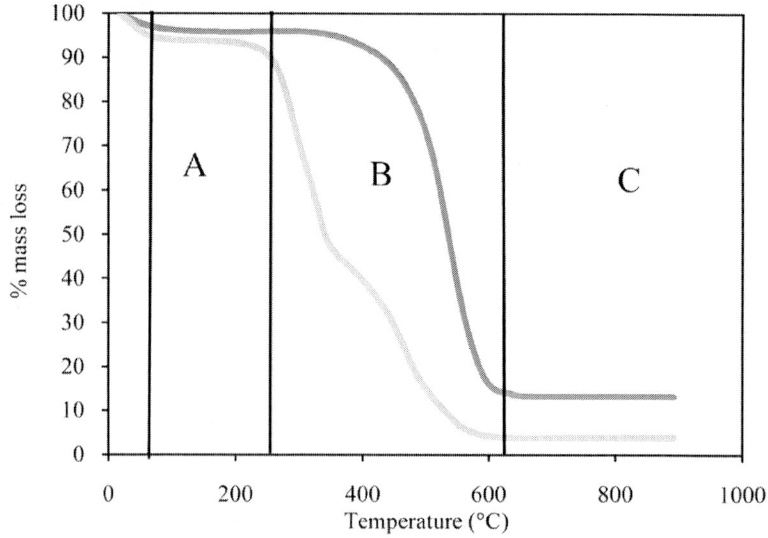


**Figure 3.** TG plots of coal and eucalyptus.

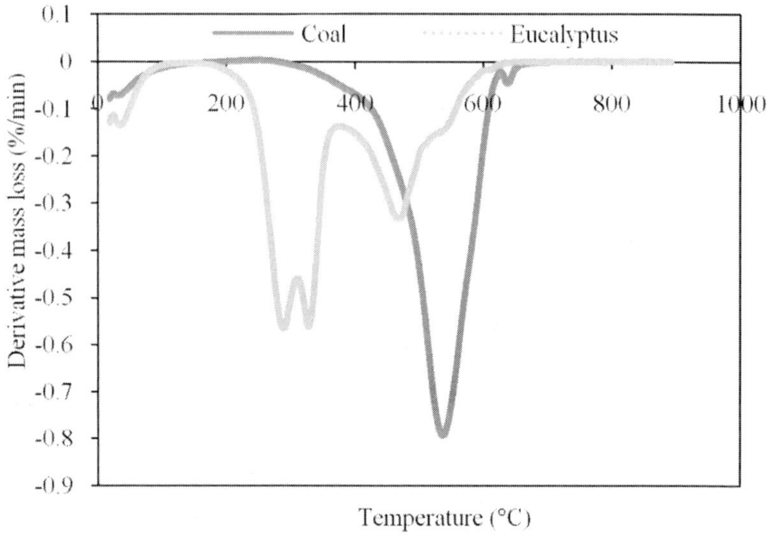


**Figure 4**. DTG plots of coal and eucalyptus.

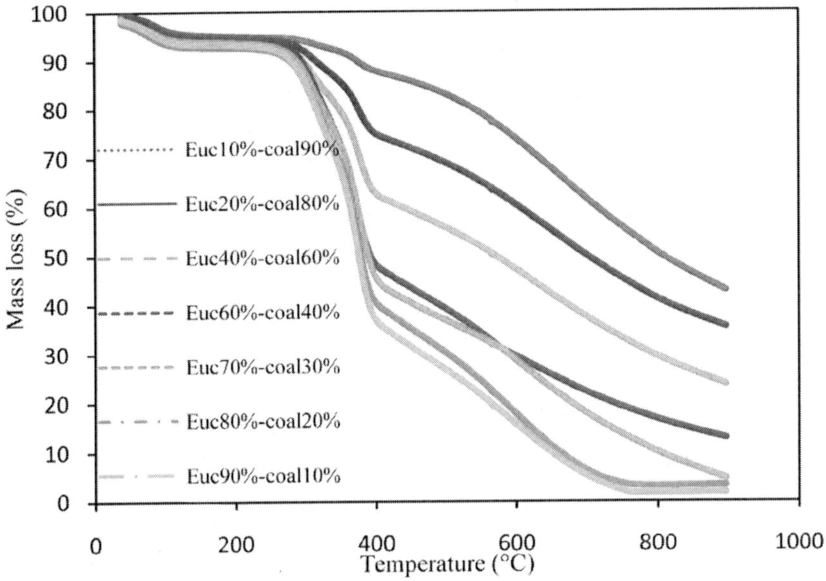


**Figure 5.** TG plots of the eucalyptus/coal blends.

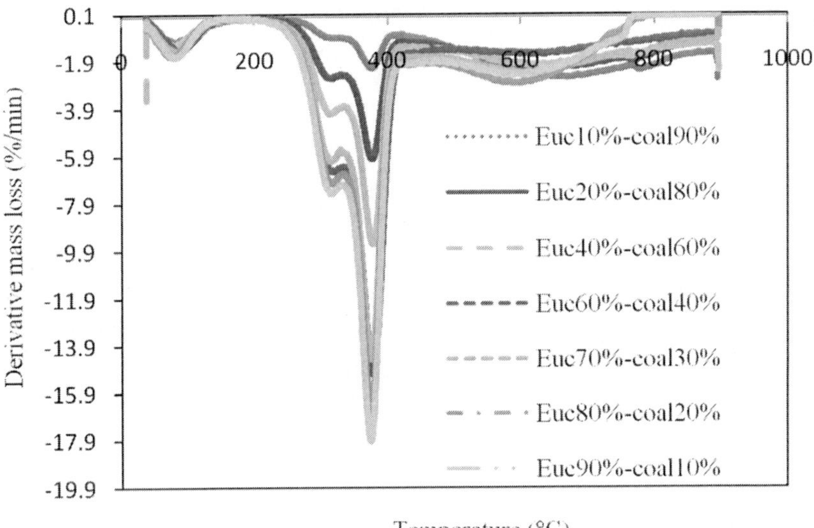


**Figure 6.** DTG plots of the eucalyptus/coal blends.

It is evident from Figure 5 that the addition of eucalyptus results in higher rate of devolatilization of the blend even at lower temperatures, the more the eucalyptus added to the blend the faster the reaction. The moisture content was observed to be almost the same for all the blends and the differences are clear in stage B where volatile matter is released and rapid devolatilization is taking place. This proves that there is a synergistic effect between coal and eucalyptus with eucalyptus acting as a catalyst for coal degradation due to its high volatile matter content, and the amount of ash left decreasing with the increase in the addition of eucalyptus. This therefore lowers the gasification temperature of the coal/biomass blend. Even Figure 6 shows that the $euc_{90\%}$-$coal_{10\%}$ blend had the fastest degradation rate and $euc_{10\%}$-$coal_{90\%}$ had the slowest as previously stated, than the devolatilization of coal alone.

### 3.5. ANOVA One Way Test

Analysis of variance (ANOVA) is a general method that is used for studying sampled-data relationships. This method permits the difference between two or more sample means to be analyzed. Its purpose is to test for significant differences between means, and this is done by analyzing the variances. ANOVA was conducted on the TGA data. Table 4 shows the summary of the ANOVA one way test.

The p value was found to be $4.8477 \times 10^{-253}$, which is far less than 0.05, therefore there is a significant difference between coal, eucalyptus and their blends, in terms of thermal degradation.

## 4. CONCLUSION

This study investigated the properties of the eucalyptus/coal blends and their suitability for co-gasification, with reference to pure eucalyptus and pure coal. The results suggested that blending biomass with coal result in an efficient clean energy with minimal negative impacts in the gasifier during co-gasification. Also, all the charac terization techniques suggested a strong existence of synergy between biomass and coal with biomass having a catalytic effect on coal during co-gasification, pyrolysis or combustion. The trace elements such as magnesium, aluminium, silicates and sulphur are then diluted when higher quantities of eucalyptus are present in the blends and this would reduce the risk of slagging during gasification of the material, which could be a challenge when coal is gasified alone.

## ACKNOWLEDGEMENTS

This research was supported by the Fort Hare Institute of Technology and Eskom.

## REFERENCES

1. Garcia, R., Pizarro, C., Lavin, A.G. and Bueno, J.L. (2012) Characterization of Spanish Biomass Wastes for Energy Use. Bioresource Technology, 103, 249-258.

2. Long, H.A. and Wang, T. (2011) Case Studies for Biomass/Coal Co-Gasification in IGCC Applications. Proceedings of ASME Turbo Expo, Vancouver.

3. Shen, L.H., Gao, Y. and Xiao, J. (2008) Simulation of Hydrogen Production from Biomass Gasification in Interconnected Fluidized Beds. Biomass and Bioenergy, 32, 120-127.

4. Brar J.S., Singh, K., Wang, J. and Kumar, S. (2012) Co-Gasification of Coal and Biomass—A Review. International Journal of Forestry Research.

5. Pan, Y.G., Velo, E. and Puigjaner, L. (1996) Pyrolysis of Blends of Biomass with Poor Coals. Fuel, 75, 412-418.

6. Mamphweli, N.S. and Meyer, E.L. (2009) Implementation of the Biomass Gasification Project for Community Empowerment at Melani Village, Eastern Cape, South Africa. Renewable Energy, 34, 2923-2927.

7. Ayhan, D. (2004) Combustion Characteristics of Different Biomass Fuels. Progress in Energy and Combustion Science, 30, 219-230.

8. Kumabe, K., Hanaoka, T., Fujimoto, S., Minowa, T. and Sakanishi, K. (2007) Co-Gasification of Woody Biomass and Coal with Air and Steam. Fuel, 86, 684-689.

# Chapter 13

# Analysis on the Governing Reactions in Coal Oxidation at Temperatures up to 400˚C

**Jing Zhan, Haihui Wang*, Feng Zhu, Shengnan Song**

*State Key Laboratory of Fire Science, University of Science and Technology of China, Hefei, China*

## ABSTRACT

The present study aims to further understanding of the principal reactions that occur during coal oxidation at moderate temperatures. Mass change and heat evolution of a sample were monitored by thermo-gravimetric analysis coupled with differential thermal analysis (TGA/DTA). Gaseous and solid products were traced using online or in situ Fourier trans- form infrared spectroscopy (FTIR). Measurements were conducted by heating the samples up to 400˚C, with the $O_2$ concentration in the reaction medium set at 0, 10, 21, and 40 vol%, respectively. It was observed that the mass increase of a sample between 150˚C and ~275˚C was a result of the accumulation of C=O containing species in the coal structure, whereas substantial mass loss and heat evolution of a sample at ~400˚C can be attributed to the significant involvement of the direct "burn-off" reaction. Enrichment of $O_2$ in the reaction medium leads to the acceleration in oxygen chemisorption, formation and decomposition of the solid oxygenated complexes, as well as the "burn-off" reaction. With the temperature increasing, the oxidation process governed by oxygen chemisorption gradually shifts to that by significant decomposition reactions, and eventually to that by the direct "burn-off" reaction. Temperature boundaries of these stages can be determined using parameters defined based on a set of TG/DTA data. Shift in the governing reactions is essentially due to the diverse requirements of reactants of the reactions and their energy barriers to be overcome. In en- gineering practice, the phenomena of self-heating and spontaneous combustion of coal correspond to chemisorption and the direct "burn-off" reaction, respectively.

**Keywords:** Coal; Oxidation at Moderate Temperatures; Reaction Product; Governing Reaction; Self-Heating; Spontaneous Combustion

## 1. INTRODUCTION

Coal oxidation at moderate temperatures (≤400˚C) alters the coal properties as fuel, thereby making pronounced impact on its subsequent utilization [1] [2] . As an exo- thermic phenomenon, coal oxidation has direct link to self-heating and spontaneous combustion of coal, which poses fire hazard in coal industries. Up to date, enormous efforts have been made to understand the chemistry and heat effects of the oxidation process [1] - [4] .

Coal oxidation at moderate temperatures is a complicated process affected by several factors, including coal compositions, temperature and oxygen concentration in the environment [1] - [3] . As stated in the literature [2] [3] the rate of oxygen consumption almost doubles by every increment of 10˚C at temperatures below 90˚C. At a spe- cific temperature, an oxygen-enriched environment may result in acceleration in the oxidation rate, thereby pro- ducing a larger amount of heat during this process.

In the past two decades, typical research work on the effect of temperature on coal oxidation can be attributed to the observations based on the thermo-gravimetric ana- lysis (TGA) and associated techniques. Crelling and co- workers [5] monitored the temperature-dependent mass change of a series of coals oxidized in the aerial atmosphere, and then suggested several temperature parameters to describe the profile of the TG curves collected. The parameters below 400˚C include "temperature of initial chemisorptions" $T_1$, "temperature of maximum rate of chemisorptions" $T_2$, and "temperature of combustion on- set" $T_3$. Based on these temperature parameters, He and Wang [6] proposed four stages of the coal oxidation at low and moderate temperatures, i.e., moisture vaporiza- tion, mass increase by absorbing oxygen, degradation caused by heating, and combustion. The rate of mass in- crease at the second stage was defined and then utilized to evaluate the propensity of various coals towards spon- taneous combustion [6] [7] .

So far, significant progress has been made in under- standing the mechanism of coal oxidation. It has been identified that two independent reaction sequences pro- ceed during the oxidation process, and one of them is the chemisorption sequence [2] [4] [8] . Chemisorption se- quence consists of several steps, including the chemisorption of oxygen, as well as formation and decompose- tion of unstable carbon-oxygen complexes (intermediates). The other reaction sequence is the so-called direct "burn- off" reaction, which occurs at specific sites in the coal structure. Given that the intermediate process is signifi- cantly rapid, direct "burn-off" reaction is described as a simultaneous formation of the gas products, i.e., $CO_2$, CO, and $H_2O$. These two reaction sequences are briefly depicted as follows:

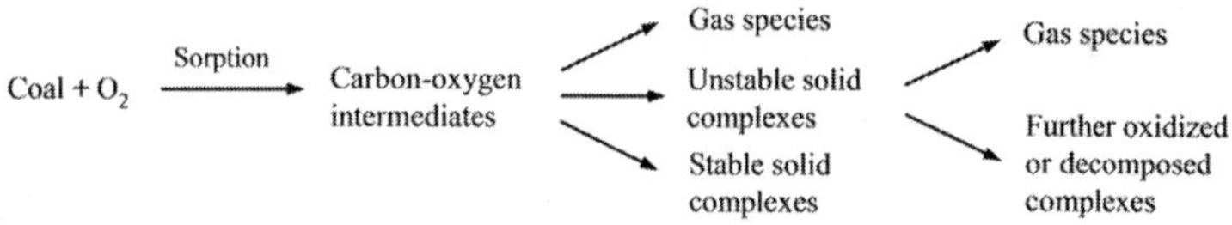


(1)

Received 17 March 2014; revised 21 April 2014; accepted 23 May 2014

$$\text{Coal} + O_2 \xrightarrow{\text{'Burn-off'}} CO_2, CO \text{ and } H_2O \tag{2}$$

Although the occurrence of the direct "burn-off" reac- tion at low temperatures (<150˚C) has been verified by the existing measurements and theoretical analysis [2] [4] [8] [9] , the dependence of this reaction to temperature and its relationship with the chemisorption sequence have not been clearly elucidated. In addition, extensive measurements have been conducted to verify the effect of $O_2$ concentration on coal oxidation; however, no rigo- rous efforts have been made to understand the role of $O_2$ molecules in altering the multi-step reactions that occur during the oxidation process.

The present study aims to further understanding of the principal reactions that occur during coal oxidation at moderate temperatures and lay a methodical basis for developing new techniques to characterize the coals and to assess their propensity towards self-heating and spon- taneous combustion. Using TGA and differential thermal analysis (DTA) coupling technique, the mass change and heat evolution are measured during coal oxidation un- der various oxygen concentrations and at temperatures up to 400˚C. Gas and solid products formed on coal surfaces are also monitored by means of a TGA in connection with a Fourier transform infrared spectrometer (TGA/ FTIR), as well as an in-situ FTIR technique. The nature of different reactions and their inherent relationships are then explored based on a series of experimental results obtained.

## 2. EXPERIMENTAL

A bituminous coal was collected from Yanzhou Colliery, Shandong Province. The coal was crushed and ground into particles. Size distribution of the particles was meas- ured using laser particle-size-distribution tester JL-1177, and the volumetric mean diameter of the particles (D [v, 0.5]) was found to be ~1.6 μm. Fine coal particles were used to eliminate the possible reaction regime controlled by gas diffusion in coal pores during the oxidation pro- cess [2] [10] . Proximate and ultimate analyses of the coal are given in Table 1.

A TGA/DTA instrument was used to evaluate pyroly- sis and oxidation behaviors of raw coal samples, and the one adopted was a Shimadzu DTG-60H thermal analysis system with a nominal resolution of mass of 0.1 μg. DTA signals at a sensitivity of 0.1 μV were directly converted into heat flow rates in the units of mW per sample cur- rent mass using a manufacturer-provided software. A sample of ~10 mg in mass was quickly and loosely placed on the aluminum crucible. The sample tempera- ture was then raised from room temperature to 400°C at a heating rate of 4°C/min. Such a low heating rate employed was to reduce the thermal lag of a sample during heating.

**Table 1.** Proximate and ultimate analyses for the coal used in the current study.

| Proximate analysis (air-dry basis) | | Ultimate analysis (dry ash free basis) | |
|---|---|---|---|
| Moisture (%) | 2.8 | C (%) | 83.5 |
| Ash (%) | 13.1 | H (%) | 5.0 |
| Volatile matter (%) | 32.5 | N (%) | 1.2 |
| Fixed carbon (%) | 51.1 | S (%) | 0.6 |
| Total sulfur (%) | 0.5 | O (%) | 9.7 |

Flow rate of the atmosphere purging into the furnace was set at 50 $cm^3$/min. Industrial-grade nitrogen gas was used in the pyrolysis experiment, and the oxidation mea- surements of the samples were performed in atmospheres that contain $O_2$ at 10, 21, and 40 vol%. The experimental uncertainties for mass change were £0.01 mg.

Coupled with a TGA (TA Q5000IR) instrument, an online FTIR spectrometer was used to trace the effluent gas products of a sample. During the measurement, the gas line was maintained at 230˚C, and the spectra were picked up in the range between 4000 and 400 $cm^{-1}$. Solid products on coal surfaces were traced using an in-situ FTIR spectrometer, equipped with a high temperature reaction chamber, a Bruker IFS 66V/S spectrometer, and a diffuse reflectance accessory (Praying Mantis™, Har- rick). A sample was placed in the reaction chamber, and the gas medium with specific $O_2$ concentration was then introduced into the chamber slowly. A measurement commenced after 30 min of gas environment preparation. The spectra of the solid surface at various temperatures were collected between 4000 and 700 $cm^{-1}$. These two types of FTIR measurements were both started by heat- ing the samples from the ambient temperature to 400˚C at 4˚C/min. During the operation of the spectrometers, the scanning times for both the background and sample were set to 32 at a resolution of 4 $cm^{-1}$. Differential spec- tra were generated by means of the software Omnic 7.4 (Thermo Fisher Scientific).

## 3. RESULTS AND DISCUSSION

### 3.1. Phenomena Observed by Analytical Instrumentations

The TG/DTA data for the samples either under $N_2$ or aerial atmosphere are shown in Figure 1. As shown by the TG results collected under $N_2$ atmosphere, the sample mass undergoes progressive decrease with the tempera- ture increase. The mass loss becomes significant at tem- peratures > 350˚C. Negative values of the DTA signals were obtained at the beginning of the measurement, but immediately returned to zero and remained constant throughout the measurement. Under aerial atmosphere, the sample undergoes a slight decline in its mass, which corresponds to the endothermic stage of the DTA signals. Afterwards, the sample mass displays progressive in- crease, and then reaches the maximum at ~275°C; meanwhile, the DTA data become positive, and the rate of heat flow reaches its maximum at ~300°C. As the temperature increases further, the sample mass begins to decrease rapidly, whereas the rate of heat flow exhibits a decline followed by a considerable increase. The trends displayed by the DTA curve for the sample under the aerial atmosphere are consistent with those reported by Banerjee and co-worker [11] .

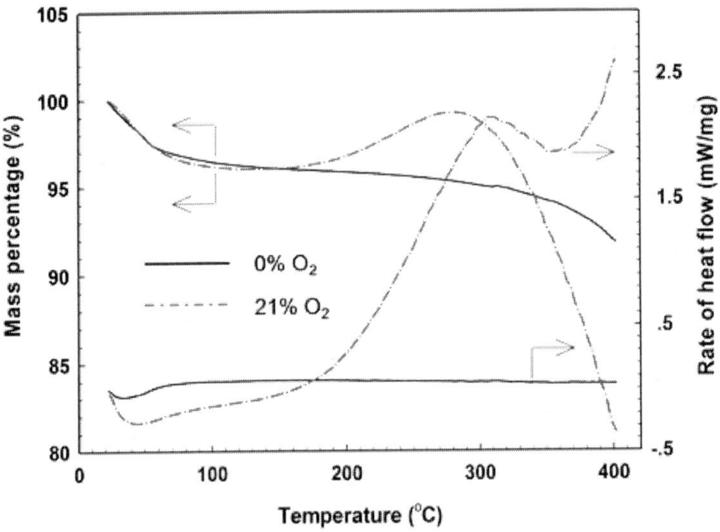


**Figure 1.** TG and DTA curves obtained by heating the samples either in nitrogen or air at a rate of 4°C/min.

During pyrolysis of a sample in nitrogen, the mass loss at the initial stage is mainly ascribed to the removal of moisture. The progressive mass change between 150°C and 300°C indicates a low rate of decomposition reactions. At ~350°C, the decomposition reactions begin to accele- rate, resulting in significant mass loss. For the sample under oxidation in air, the initial mass loss was also mainly led by water evaporation, which is demonstrated by the negative values in the rate of heat flow. As a result of oxygen uptake and formation of solid oxygenated complexes, the sample mass starts to increase, and the corresponding rate of heat flow becomes positive. After reaching the maximum at ~275˚C, the sample mass tends to decrease with the further increase in temperature. This is a sign to indicate the acceleration in the rate of thermal decomposition (endothermic) reactions: on one hand, it makes the sample mass decrease; on the other hand, the heat absorbed by the thermal decomposition of unstable oxygenated complexes offsets in part the heat evolved by the oxygen chemisorption, thereby reducing the rate of heat flow. The DTA curve rebounds again after slight decrease, which is assigned to the significant mass loss shown in the TG curve. This phenomenon suggests that another exothermic reaction comes into action at this stage.

The variations of the TG and DTA curves for the sam- ple heated in aerial atmosphere are essentially related to the multiple reactions that occur during the oxidation process. By contrast, the sample heated under nitrogen atmosphere always undergoes thermal decomposition reactions throughout the measurement. The obvious dif- ferences shown by these two sets of TG/DTA curves in- dicate that the reaction medium plays an important role in altering the behavior of a coal sample during heating.

Li et al. [12] reported their TG/DTA results for oxida- tion of a bituminous coal and several types of refuse- derived fuels. The trends of mass change and the cumulative heat for all of the samples observed in the temper- ature range between 0˚C and 400˚C are similar to those described in the present study. Instead of providing de- tailed analysis on the experimental data, they stated that the progressive increase in mass and the rate of heat flow at temperatures > 120˚C is due to the reactions induced by oxygen adsorption.

The TG curve obtained under aerial atmosphere was analyzed using the parameters defined by Crelling et al. [5] , with three characteristic temperatures found below 400˚C. Given the fluctuation of the derivative TG (DTG) data at temperatures between 120˚C and 170˚C, the pa- rameter $T_1$ falls between 130.1˚C and 168.0˚C, whereas the parameters $T_2$ and $T_3$ are evaluated at 238.4˚C and 276.2˚C, respectively. $T_3$ cannot be considered as the "onset temperature of combustion", because the corres- ponding DTA data did not illustrate a drastic increase at temperatures > $T_3$. Direct scanning electron microscope (SEM) images also show no sign of combustion for the sample even after heating to 400˚C in air.

Gaseous species formed during the coal oxidation process were traced using an online FTIR spectrometer; the absorption intensities observed at various tempera- tures are reported in Figure 2. It is evident that within the temperature range studied, the gaseous species released are $CO_2$ (2400 cm$^{-1}$ - 2220 cm$^{-1}$), CO (2220 cm$^{-1}$ - 2050 cm$^{-1}$), and $H_2O$ (1800 cm$^{-1}$ - 1400 cm$^{-1}$) [13] [14] . The absorption peak of $CO_2$ becomes obvious at ~200˚C, and its intensity increases rapidly with temperature in- crease. As the major gas product, $CO_2$ formation is di- rectly responsible for the sample mass loss at a higher temperature during the TGA measurement.

FTIR spectra of the solid species obtained at various temperatures during the sample oxidation in air are shown in Figure 3. The major absorption peaks and their cor- responding species (organic functional groups) were iden- tified and labeled. The observed species usually contain O-H, C-H, C=C, C-O, and C=O bonds [15] [16] . Chan- ges in the functional groups on the sample surfaces were further examined by subtracting FTIR spectra collected at various temperatures (Figure 4). The curve a in Figure 4, which is the subtracted FTIR spectra for a sample at room temperature from those obtained after heating to 275˚C in nitrogen, indicates that a coal sample mainly experiences the water removal and hydroxyl decomposi- tion in an inert atmosphere [14] [16] .

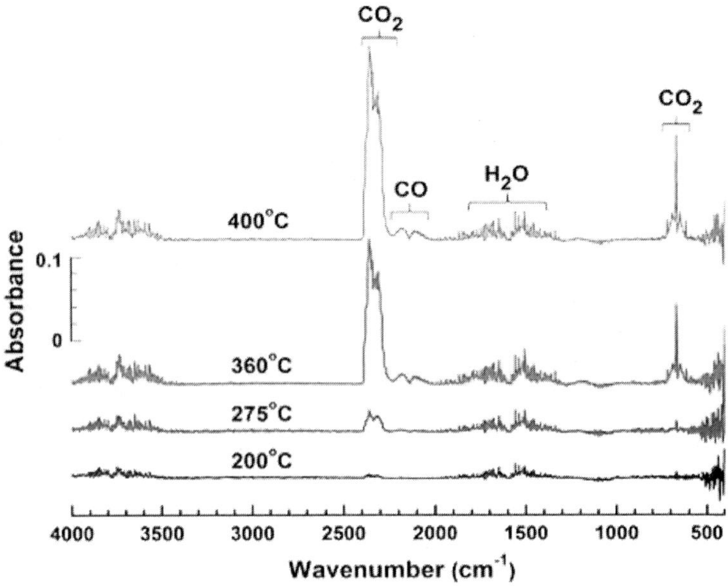


**Figure 2.** FTIR spectra of gas products evolved at various temperatures for a sample heated at 4˚C/min in air.

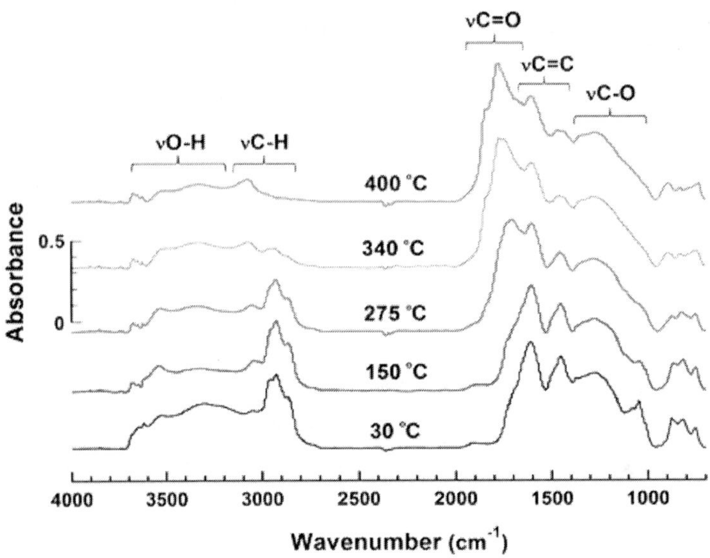


**Figure 3.** FTIR spectra observed on the surfaces of a sample at various temperatures during heating at 4˚C/min in air.

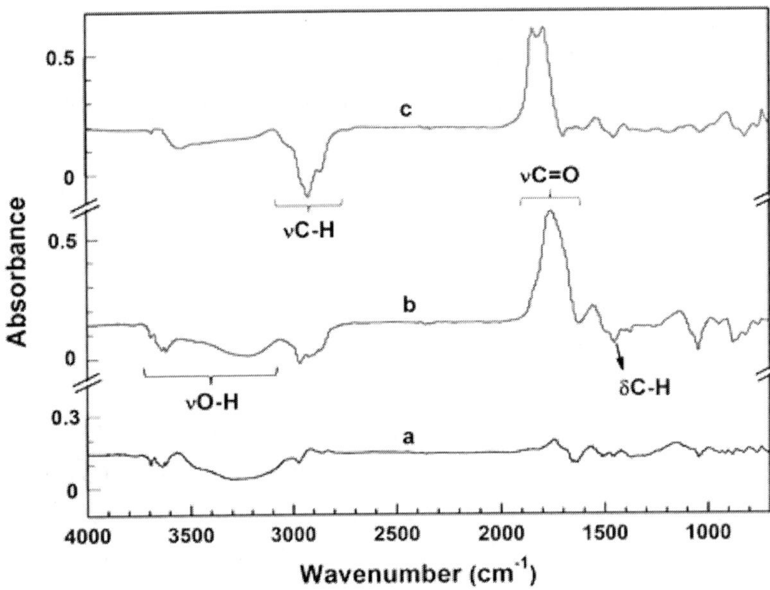


**Figure 4.** Comparisons of the subtracted FTIR spectra for the samples at ambient temperature from those after heating to 275˚C either in nitrogen (a) or in air (b), as well as the spectra for the sample oxidized to 275˚C from those after heating to 400˚C in air (c).

Significant changes have been observed in the major absorption bands for the sample oxidized at 275˚C (curve b in Figure 4). The weakening of the bands in the region 3680 $cm^{-1}$ - 3100 $cm^{-1}$ is also due to water removal and decomposition of hydroxyls. The intensities of the peaks for vC-H (v: stretching vibration) (3050 $cm^{-1}$ - 2800 $cm^{-1}$) and δC-H (δ: bending vibration) (1450 $cm^{-1}$) decrease significantly, indicating loss of hydrocarbon species un- der the aerial atmosphere [4] [17] . Strong positive ab- sorption peaks centered at ~1760 $cm^{-1}$ appear, implying the formation of abundant oxygenated complexes con- taining C=O bonds. It is suggested that the oxidation products formed at ~275˚C are mainly those containing C=O bonds, including aldehydes and carboxylic acids.

The subtracted FTIR spectra for a sample oxidized at 275˚C from those after heating to 400˚C in air are shown as curve c in Figure 4. The appearance of the negative absorption bands at 3630 $cm^{-1}$ - 3100 $cm^{-1}$ for νO-H demonstrates that with the further increase in temperature, hydroxyls continue to be condensed or oxidized. Similar to curve b in Figure 4, the negative bands assigned to νC-H (3100 $cm^{-1}$ - 2800 $cm^{-1}$) and δC-H (1450 $cm^{-1}$) are still strong, thereby indicating accelerated rate of oxida- tion reactions with alkyls. The strong positive absorption peaks of C=O at higher bands (2000 $cm^{-1}$ - 1720 $cm^{-1}$) mainly belong to the anhydrides and esters [4] [15] . Weak negative absorption bands of C=O appear in the region of 1720 $cm^{-1}$ - 1670 $cm^{-1}$, corresponding to alde- hydes and carboxylic acids. At higher temperatures, more anhydrides and esters are formed as a result of the oxida- tion of alkyls and the decomposition (or condensation) reactions of carboxylic acids [15] [17] . In addition, the decomposition and condensation reactions among the existing complexes containing C=O bonds may cause a loss of certain C=O bonds [4] [18] .

### 3.2. Effect of Oxygen Concentration on the Coal Oxidation

The TG curves for coal samples oxidized at different oxygen concentrations are compared in Figure 5(a).

These curves have similar variation patterns, which al- most overlap at the initial stage. With the increase in oxygen concentration, the mass increase peak rises, whereas the peak temperature decreases. In the higher temperature region, an increase in oxygen concentration results in more significant decrease in the TG curve, in- dicating acceleration in the mass loss rate of a sample. The residual mass at the conclusion of the measurements decreases by ~7 wt% with an increase of $O_2$ concentra- tion from 10 to 40 vol%.

As shown in Figure 5(b), the water evaporation pro- cess contributes to the variation in the DTA curves at the initial stage. At the stage of mass increase of a sample, the rate of heat flow increases more significantly with the increase in $O_2$ concentration, which corresponds to an increased amount of heat evolved. The rate of heat flow decreases slightly at a temperature exceeding ~300˚C, but rebounds at ~360˚C. The rate increment between ~360˚C and 400˚C consistently increases with the increa- se in $O_2$ concentration.

Several parameters are defined to characterize these TG and DTA curves, including three characteristic tem- peratures, the percentage of mass increase at ~275˚C, the amount of heat evolved in association with the sample mass increase, and the percentage of increase in the heat flow rate at 400˚C. The parameter $T_{c1}$ is the temperature of the complete water evaporation, i.e. the starting point at which the rate of heat flow becomes positive. $T_{c2}$ is the peak temperature at the mass increase stage, whereas $T_{c3}$ is the second transition temperature of the exothermic peak in the DTA curve (Figure 5(a) & Figure 5(b)). The per- centage of mass increase at ~275˚C is determined by sub- tracting the minimum sample mass at ~150˚C from the maximum mass between 150˚C and 300˚C, and divided by the minimum sample mass at ~150˚C [16] . The amount of heat evolved in association with the sample mass in- crease is obtained by integrating the exothermic peak area of the DTA curve between $T_{c1}$ and $T_{c3}$ (Figure 5(b)).

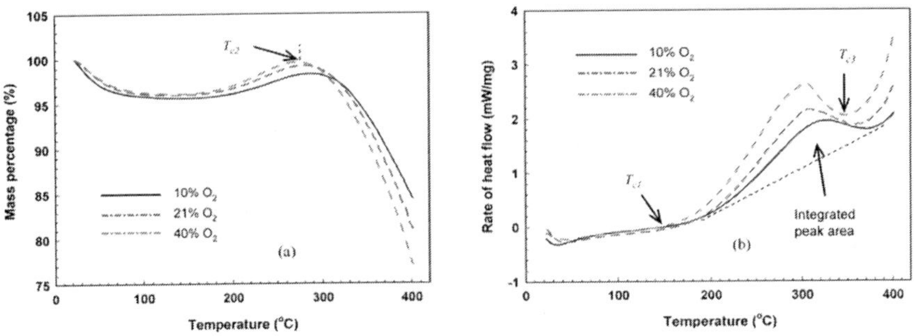


**Figure 5.** TG (a) and DTA (b) results measured for coal samples oxidized under different oxygen concentrations.

The percentage of increase in the heat flow rate at 400˚C is defined by the heat flow rate at 400˚C minus that at $T_{c3}$ and then divided by that at $T_{c3}$. The results of these pa- rameters are given in Table 2.

As shown in Table 2, with the increase in oxygen concentration, $T_{c2}$ and $T_{c3}$ decrease, but the percentage of mass increase at ~275˚C, the associated amount of heat evolved, and the percentage of increase in the heat flow rate at 400˚C increase. For the oxygen concentration va- rying from 10 vol% to 21 vol%, the percentage of mass increase grows from 2.74% to 3.35% (0.61% increase). However, once the oxygen concentration varies from 21 vol% to 40 vol%, the extent of increase is only 0.22%. A continuously increasing trend is observed for the asso- ciated amount of heat evolved. These results suggest no linear dependence between the percentage of mass in- crease and the associated amount of heat evolved at va- rying oxygen concentrations. This is understandable, be- cause the mass increase and the heat release are the com- prehensive performance of a series of exothermic and endothermic reactions taking place during the oxidation process.

Temperature-dependent emission of $CO_2$ for the coal samples oxidized under different oxygen concentrations is shown in Figure 6. For a specific measurement, the absorption intensity of $CO_2$ varies in consistency with the corresponding rate of heat flow (refer to Figure 5(b)), and its variations can be split into two distinct phases. The first phase is between 200˚C and 360˚C for the sam- ple oxidized in air. With the temperature increase, the intensity of the absorption peak grows progressively, fol- lowed by a quasi-steady increase at temperatures > 300˚C. These results somehow reflect the variation trends in the rate of decomposition reactions. In the second phase ($^3$360˚C), the rate of $CO_2$ production undergoes more drastic and persistent increase, indicating the commen- cement of a different mechanism of gas evolution.

Consistency in the patterns of gas evolution at these two consecutive phases has been observed in the mea- surements at various $O_2$ concentrations. By recalling the variation trends of the DTA signals (Figure 5(b)), we confirm that at temperatures > ~360˚C, the gas products may be sourced by an extra formation mechanism, and oxygen plays a role in both promoting the rate of $CO_2$ formation and the amount of heat evolved. There is no doubt that only the direct "burn-off" reaction serves these purposes.

The compositions on coal surfaces traced during the oxidation under 40 vol% $O_2$ by using an in-situ FTIR spectrometer are compared to those under 10 vol% $O_2$ (Figure 7). Curves $a_1$ and $a_2$ are the subtracted FTIR spectra for the samples at ambient temperature from those after heating to 275°C. With the increase in $O_2$ concentration from 10 vol% to 40 vol%, the negative absorption intensity of νC-H bands (3100 $cm^{-1}$ - 2800 $cm^{-1}$) increases; meanwhile, the positive absorption inten- sity of νC=O bands (1880 $cm^{-1}$ - 1650 $cm^{-1}$) becomes stronger. These results demonstrate that with the increase in the oxygen concentration, greater amount of $O_2$ is ad- sorbed, and more hydrocarbon species are oxidized. As a result, extra amounts of aldehydes and carboxylic acids are formed in the coal structure.

**Table 2.** List of the data for $T_{c2}$ and $T_{c3}$, percentage of mass increase at ~275°C, the amount of heat evolved in association with the sample mass increase, and the percentage of increase in the heat flow rate at 400°C.

| Oxygen conc. (vol%) | $T_{c2}$ (°C) | $T_{c3}$ (°C) | Perc. of mass increase at ~275°C (%) | Amount of heat evolved (kJ/(kg raw coal)) | Perc. of increase in heat flow rate at 400°C (%) |
|---|---|---|---|---|---|
| 0 | - | - | - | - | - |
| 10 | 288.3 | 374.7 | 2.74 | 870 | 16.0 |
| 21 | 280.1 | 364.4 | 3.35 | 1040 | 38.9 |
| 40 | 272.0 | 347.7 | 3.57 | 1320 | 74.3 |

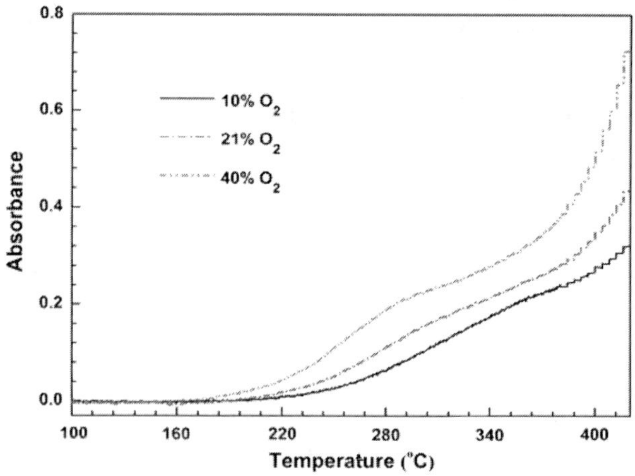


**Figure 6.** Temperature-dependent FTIR absorbance for $CO_2$ evolved from coal samples oxidized under various oxygen con- centrations.

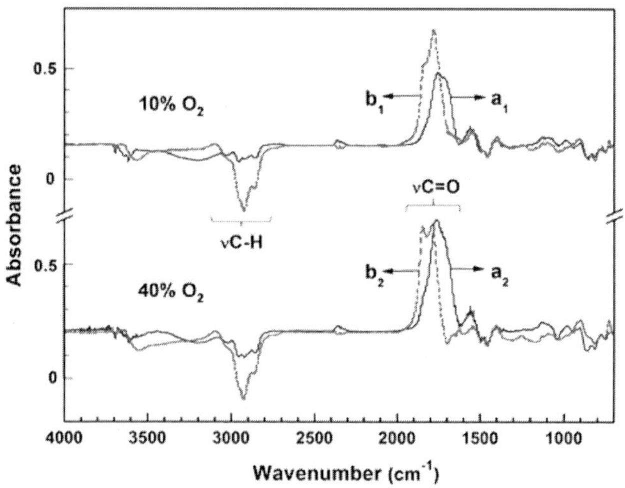


**Figure 7.** Comparisons of the subtracted FTIR spectra for the samples oxidized in the atmospheres containing 10 ($a_1$ & $b_1$) and 40 vol% $O_2$ ($a_2$ & $b_2$), respectively. Curves $a_1$ & $a_2$ were obtained by subtracting the spectra of the samples at ambient from those after heating to 275°C under 10 or 40 vol% $O_2$, and curves $b_1$ & $b_2$ were generated by subtracting the spectra of the samples oxidized to 275°C from those after heating to 400°C.

The phenomena observed above correspond to the fol- lowing reaction steps:

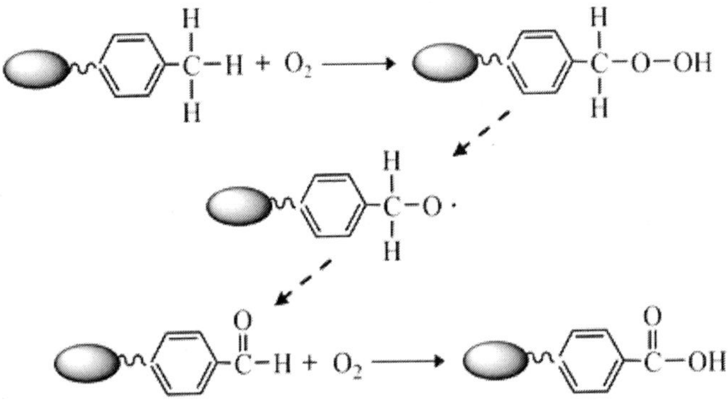

(3)

(4)

where Reaction 4 contains a few sub-steps in principal, including the activation of carbonyl species by chain carriers (H-O) and the subsequent formation of peroxides at the carbon centers [2] [4] [8] .

It is evident that an increase in the oxygen concentra- tion speeds up the rates of these two reactions. This has been reflected by an increase in the percentage of mass increase at ~275˚C. The increased rates of these reactions implies an increase in the amount of heat released, i.e. the enlarged exothermic peak area on the individual DTA curves (refer to Figure 5(b) and Table 2).

Obvious differences are also observed between the subtracted FTIR spectra for the samples oxidized at 275˚C and those after heating to 400˚C, for example, cur- ve $a_2$ versus $b_2$ in Figure 7. Stronger absorption peaks of νC-H bands are observed in the region of 3100 $cm^{-1}$ - 2800 $cm^{-1}$, implying that larger amount of hydrocarbon species is consumed at higher temperatures. Unlike the phenomena illustrated by the curves $a_1$ and $b_1$, the ab- sorption intensities of νC=O bonds are almost identical for the sample oxidized to 275˚C and 400˚C under 40 vol% $O_2$. These results indicate that at ~400˚C, more hydro- carbon species in coal structure are consumed via direct "burn-off" reaction, rather than the chemisorption se- quence. This is also an evidence to demonstrate that an increase in oxygen concentration may speed up the direct "burn-off" reaction, which explains the obvious reduc- tion in the residual mass at the end of the TGA measure- ment in an oxygen-enriched atmosphere.

Following the above statements, it is straightforward to establish that the considerable difference in mass loss at the end of the measurements for the samples oxidized in either an oxygen-leaned or oxygen-enriched environ- ment is essentially contributed by the direct "burn-off" reaction, rather than the chemisorption sequence. This happens only when a shift of the governing reactions is performed from the chemisorption sequence, which is primarily operated at temperatures below 360˚C, to the direct "burn-off" reaction at the temperatures above 360˚C.

### 3.3. Characterization of the Oxidation Process at Different Temperature Ranges

As demonstrated by the TG and DTA results, identical trends are exhibited for the samples oxidized in various oxygen concentrations (Figure 5). Comparable patterns of the mass change and heat evolution have been re- ported in the literature for various types of coals and biomass using the same experimental approach [6] [7] [12] . Thus, TG/DTA curves are used to characterize the coal oxidation process at different temperature ranges. Based on the TG/DTA curves obtained by oxidation mea- surement in air, the oxidation process can be divided into the following typical stages: 1) water evaporation ($T \leq T_{c1}$), 2) oxygen chemisorption ($T_{c1} < T \leq T_{c2}$), 3) signifi- cant decomposition reactions ($T_{c2} < T \leq T_{c3}$), and 4) the stage dominated by the direct "burn-off" reaction ($T > T_{c3}$) (Figure 8).

As is well known, coal contains certain amount of wa- ter, and during its oxidation below $T_{c1}$, the exothermic oxidation reactions and endothermic water evaporation coexist. Given that the amount of heat released by the former is smaller than that absorbed by the later, negative values are presented on the DTA curve at this stage.

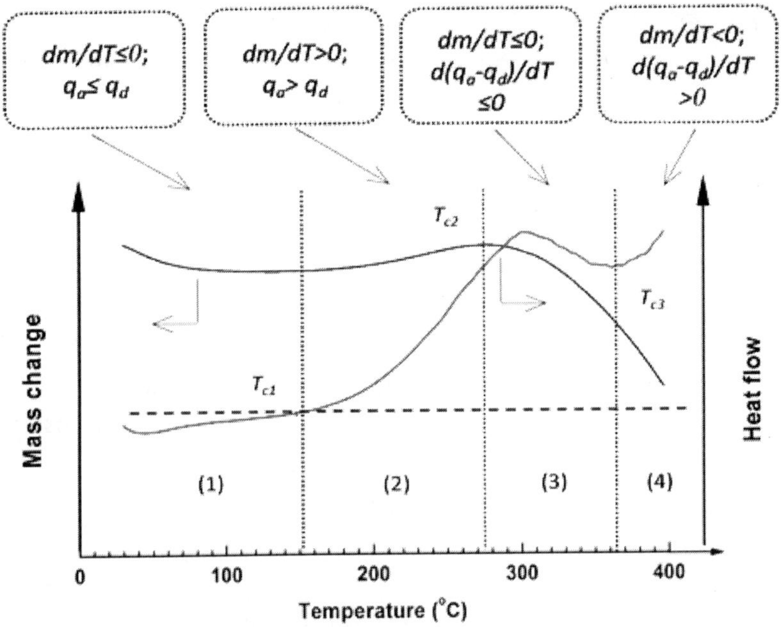


**Figure 8.** Four stages of coal oxidation up to 400˚C determined by a set of TG/DTA curves: (1) Water evaporation; (2) Oxygen chemisorption; (3) Significant decomposition reactions; and (4) The stage dominated by direct "burn-off" reaction.

Meanwhile, the mass loss led by water evaporation is much larger than the mass increase due to oxygen ad- sorption, thereby resulting in the observed decline in the TG curve.

At the stage of oxygen chemisorption, the internal coal pores begin to expose to the oxidation medium because of water removal; hence, the rate of chemisorption in- creases. The mass increase of a sample reaches the maxi- mum at ~275˚C. At this stage, the decomposition of un- stable oxygenated complexes does occur, and the amount of heat evolved is the compromised result of the instan- taneous heat of exothermic chemisorption reactions ($q_a$) and that of endothermic decomposition reactions ($q_d$). Given that thermal decomposition of unstable oxygen- nated complexes is a relatively slow process at tempera- tures between $T_{c1}$ and $T_{c2}$, the amount of heat released increases with the temperature rise, which has been dis- played by the DTA curves. The associated infrared spec- tra for solid products also clearly show that these phe- nomena are primarily contributed by the chemisorption of $O_2$ and the formation of the species containing C=O bonds, as a result of interaction of the active groups in coal structure with $O_2$.

Significant decomposition reactions occur at tempera- tures between $T_{c2}$ and $T_{c3}$, and the rate of heat evolution by oxygen adsorption is not enough to offset that con- sumed by the thermal decomposition reactions in general, i.e. $d(q_a - q_d)/dT \leq 0$. Therefore, the amount of heat re- leased $(q_a - q_d)$ reduces, which is illustrated by a decline in the rate of heat flow. Accordingly, the sample mass de- creases because of the accelerated decomposition reactions (refer to Figure 5 & Figure 6). These results show that the oxygenated complexes containing C=O bonds are relatively stable below $T_{c2}$ (<300˚C), whereas obvious decomposition occurs at temperatures ³ 300˚C, resulting in a reduction of the total amount of heat released. For this reason, $T_{c2}$ can be treated as the critical temperature at which a considerable amount of oxygenated complex- es begin to decompose. This understanding is consistent with the statements of Berkowitz [19] regarding the sta- bility of humic acids that contain such functional groups.

As the temperature increases further ($>T_{c3}$), the "burn- off" reaction is remarkably accelerated, which contri- butes to the drastic decline in the TG curves and a rapid increase in the DTA signals. Consequently, the stage of significant decomposition reactions is replaced by the stage dominated by the "burn-off" reaction. Worthy of mentioning, a coal sample at this stage is definitely not at the "real" status of combustion or burning, which has been verified by SEM observations on the sample col- lected after heating to 400˚C in air.

In the literature [5] , the characteristic temperatures are determined solely based on TG and the DTG data. As a result, these characteristic temperatures have no direct link to chemical reactions and the associated heat re- lease in coal oxidation. The definition of these characte- ristic temperatures is also questionable. Parameter $T_1$ is named as the "temperature of initial chemisorption", which is the first transition point at which the TG curve shifts from mass loss to mass gain. In fact, chemisorption starts from the beginning of a test, even though the mass gain by $O_2$ adsorption is concealed by the mass loss be- cause of water evaporation below $T_1$.

Crelling et al. [5] defined the parameter $T_2$ as the "temperature of maximum rate of chemisorptions", i.e., the first peak temperature on a DTG curve. The oxidation of coal is not a simple oxygen adsorption process, and the phenomenon of mass increase is the overall reflection of a series of reactions leading to mass gain and loss; therefore, we are unable to expect a direct link between the maximum rate of chemisorptions and the maximum rate of mass increase. The "temperature of combustion onset", $T_3$, is actually the temperature at which a sample mass reaches its maximum, which is equivalent to $T_{c\,2}$ in the present study. There is no doubt that the characte- ristic temperatures introduced in the present study have clear physical significance and solid chemistry basis.

Once the rank of coal is very high, its volatile matter content becomes low, which implies a smaller amount of C-H bonds stored in the coal structure and fewer carbon centers available for the oxygen chemisorption reactions. As a result, the stage of significant decomposition reac- tions tends to be limited, which can be seen from the typical DTA or differential scanning calorimeter (DSC) results for the semi-anthracite or anthracite [20] [21] . As the carbon content is maintained at ~90%, the decline segment on the DSC curve (at the temperatures usually £$T_{c3}$) may disappear. Under this circumstance, the stage of significant decomposition reactions is then replaced by that dominated by the direct "burn-off" reaction.

The consistent decrease in the values of $T_{c2}$ and $T_{c3}$ for the samples oxidized under the oxygen concentrations from 10 to 40 vol% is an indication of the role of $O_2$ mo- lecules in altering the oxidation process.

The presence of $O_2$ triggers the formation of the transient intermediates (Reaction 3) and their further oxidation into aldehydes and other oxygenated complexes (e.g., Reaction 4). An increase in oxygen concentration implies the accelera- tion in the rates of oxygen chemisorption and formation of solid oxygenated complexes, leading to a growth in both the percentage of mass increase at ~275˚C and the associated amount of heat evolved. In addition, enrich- ment of $O_2$ in the reaction medium indirectly intensifies the decomposition reactions of the oxygenated com- plexes and speeds up the "burn-off" reaction. As a result, the corresponding decrease in characteristic tempera- tures $T_{c2}$ and $T_{c3}$ was observed.

Shift in the governing reactions at various tempera- tures is an overall reflection of the temperature effect on these reactions, which are particularly due to their di- verse requirements of reactants and the energy barriers to overcome. Desorption of water requires overcoming the minimum energy barriers and makes it available for $O_2$ molecules to access the unveiled active carbon centers in coal pores [22] . Because of the lower energy barriers for oxygen adsorption (both physical and chemical), chemi- sorption of oxygen proceeds relatively easily compared with the direct "burn-off" reaction, especially at low tem- peratures [2] [8] . High temperatures (>360˚C) and the deformed coal structure, which is a consequence of the significant decomposition reactions, provide both energy and material basis for rapid progress in the direct "burn- off" reaction.

In engineering practice, coal may not be always ex- posed to a flowing environment, and when water is re- moved to a certain extent, the remaining water could act as a catalyst to the oxygen chemisorption reactions [2] . Oxygen chemisorption reactions provide initial energy for the self-heating of coal in the absence of external heat sources. When the temperature is high enough, the coal eventually moves toward spontaneous combustion be- cause of the significant involvement of the direct "burn- off" reaction. The self-heating and spontaneous combus- tion phenomena of coal not only correspond to two dis- tinct stages in the oxidation process, but also are con- trolled by these two major reactions, respectively. From this standpoint, in the assessment on the propensity of a coal towards self-heating and spontaneous combustion, appropriate methods should be used to focus on either the oxygen adsorption capacity or the capability of coal to participate in the direct "burn-off" reaction.

## 4. CONCLUDING REMARKS

During the oxidation at <~275˚C, the alkyl chains in coal structure are converted into C=O containing species, leading to the increase in sample mass and the associated amount of heat evolution. At temperatures ~400˚C, the direct "burn-off" reaction of hydrocarbon species, which is responsible for the higher rate of $CO_2$ emission of a sample, becomes dominated.

An increase in oxygen concentration in the reaction medium results in an increased amount of heat evolved and a decreased temperature threshold for initiating sig- nificant thermal decomposition reactions. The presence of sufficient oxygen speeds up the oxidation process by primarily promoting oxygen chemisorption, formation and decomposition of the solid complexes containing C=O groups, as well as the direct "burn-off" reaction.

The products both in gas and solid phase, as well as the heat evolution allow the coal to demonstrate distinct characteristics during its oxidation at varying tempera- tures. With the temperature increase, the oxidation pro- cess can be classified into four typical stages, namely, water evaporation, oxygen chemisorption, significant decomposition reactions, and the stage dominated by the direct "burn-off" reaction. Temperature boundaries of these stages can be determined using the parameters de- fined based on a set of TG/DTA data.

## 5. ACKNOWLEDGEMENTS

In situ FTIR measurements on solid surfaces were per- formed at the infrared and far-infrared beamline station of National Synchrotron Radiation Laboratory (NSRL) in Hefei, China. We would like to thank Assoc. Prof. Zeming Qi and his colleagues for providing assistance during the measurements.


## REFERENCES


1. van Krevelen, D.W. (1993) Coal: Typology—Chemistry—Physics—Constitution. 3rd Edition, Elsevier, New York, 627-658.

2. Wang, H.-H., Dlugogorski, B.Z. and Kennedy, E.M. (2003) Coal Oxidation at Low Temperatures: Oxygen Consumption, Oxidation Products, Reaction Mechanism and Kinetic Modeling. Progress in Energy and Combus- tion Science, 29, 487-513.

3. Carpenter, D.L. and Giddings, D.G. (1964) The Initial Stages of the Oxidation of Coal with Molecular Oxygen. I. Effect of Time, Temperature and Coal Rank on Rate of Oxygen Consumption. Fuel, 43, 247-266.

4. Clemens, A.H., Matheson, T.W. and Rogers, D.E. (1991) Low Temperature Oxidation Studies of Dried New Zeal- and Coals. Fuel, 70, 215-221.

5. Crelling, J.C., Hippo, E.J., Woerner, B.A. and West Jr., D.P. (1992) Combustion Characteristics of Selected Whole Coals and Macerals. Fuel, 71, 151-158. http://dx.doi.org/10.1016/0016-2361(92)90003-7

6. He, Q.-L. and Wang, D.-M. (2005) Comprehensive Study on Coal Oxidation Process by TG-DTA-FTIR Techniques. Journal of China Coal Society, 30, 53-57. (in Chinese)

7. Wang, J.-R., Deng, C.-B., Shan, Y.-F., Hong, L. and Lu, W.-D. (2008) A New Method for Classifying Tendency of Coals to Spontaneous Combustion. Journal of China Coal Society, 33, 47-50. (in Chinese)

8. Wang, H.-H., Dlugogorski, B.Z. and Kennedy, E.M. (2003) Analysis of the Mechanism of the Low-Tempera- ture Oxidation of Coal. Combustion and Flame, 134, 107- 117.

9. Krishnaswamy, S.K., Gunn, R.D. and Agarwal, P.K. (1996) Low-Temperature Oxidation of Coal. 2. An Expe- rimental and Modeling Investigation Using a Fixed-Bed Isothermal Flow Reactor. Fuel, 75, 344-352.

10. Wang, H.-H., Dlugogorski, B.Z. and Kennedy, E.M. (1999) Theoretical Analysis of Reaction Regimes in Low- Temperature Oxidation of Coal. Fuel, 78, 1073-1081.

11. Banerjee, S.C. and Chakravorty, R.N. (1967) Use of DTA in the Study of Spontaneous Combustion of Coal. Journal of Mines, Metals and Fuels, 15, 1-5.

12. Li, X.-R., Lim, W.-S., Yusaku, I. and Hiroshi, K. (2009) Safety Evaluation of Sewage-Sludge-Derived Fuels by Comparison with Other Fuels. Fire and Materials, 33, 187-200.

13. Carangelo, R.M., Solomon, P.R. and Gerson, D.J. (1987) Application of TG-FT i.r. to Study Hydrocarbon Structure and Kinetics. Fuel, 66, 960-967.

14. MacPhee, J.A., Giroux, L., Charland, J.-P., Gransden, J.F. and Price, J.T. (2004) Detection of Natural Oxidation of Coking Coal by TG-FTIR—Mechanistic Implications. Fuel, 83, 1855-1860.

15. Rose, H.R., Smith, D.R. and Vassallo, A.M. (1998) Study of the Oxidation of Oil Shale and Kerogen by Fourier Transform Infrared Emission Spectroscopy. Energy and Fuels, 12, 682-688.

16. Zhan, J., Wang, H.-H., Song, S.-N., Hu, Y. and Li, J. (2011) Role of an Additive in Retarding Coal Oxidation at Moderate Temperatures. Proceedings of the Combus- tion Institute, 33, 2515-2522.

17. Iglesias, M.J., de la Puente, G., Fuente, E. and Pis, J.J. (1998) Compositional and Structural Changes during Aerial Oxidation of Coal and Their Relations with Tech- nological Properties. Vibrational Spectroscopy, 17, 41-52.

18. Calemma, V., Rausa, R., Margarit, R. and Girardi, E. (1988) FT-i.r. Study of Coal Oxidation at Low Tempera- ture. Fuel, 67, 764-770.

19. Berkowitz, N. (1979) An Introduction to Coal Technolo- gy. Elsevier, New York, 95-130.

20. Pis, J.J., de la Puente, G., Fuente, E., Morán, A. and Ru- biera, F. (1996) A Study of the Self-Heating of Fresh and Oxidized Coals by Differential Thermal Analysis. Ther- mochimica Acta, 279, 93-101.

21. Fangxian, L., Shizong, L. and Youzhi, C. (2009) Thermal Analysis Study of the Effect of Coal-Burning Additives on Combustion of Coals. Journal of Thermal Analysis and Calorimetry, 95, 633-638.

**22.** Wang, H.-H. (2007) Kinetic Analysis of Dehydration of a Bituminous Coal Using TGA Technique. Energy and Fu- els, 21, 3070-3075.

# Index